SECURITY AND QUALITY OF SERVICE IN AD HOC WIRELESS NETWORKS

Ensuring secure transmission and good quality of service (QoS) are key commercial concerns in ad hoc wireless networks as their application in short range devices, sensor networks, control systems, and other areas continues to develop. Focusing on practical potential solutions, this text covers security and quality of service in ad hoc wireless networks.

Starting with a review of the basic principles of ad hoc wireless networking, coverage progresses to the vulnerabilities these networks face and the requirements and solutions necessary to tackle them. QoS in relation to ad hoc networks is covered in detail, with specific attention to routing, and the basic concepts of QoS support in unicast communication, as well as recent developments in the area. There are also chapters devoted to secure routing, intrusion detection, security in WiMax networks, and trust management, the latter of which is based on principles and practice of key management in distributed networks and authentication.

This book represents the state of the art in ad hoc wireless network security and is a valuable resource for graduate students and researchers in electrical and computer engineering, as well as for practitioners in the wireless communications industry.

AMITABH MISHRA worked at Lucent Technologies (formerly Bell Labs) for 13 years before moving to Virginia Tech. He is currently with the Center for Networks and Distributed Systems, Department of Computer Science, Johns Hopkins University. He was awarded his Ph.D. in Electrical Engineering in 1985 from McGill University. A senior member of the IEEE, he has chaired the IEEE Communications Software committee, and holds several patents in the field of wireless communications.

SECURITY AND QUALITY OF SERVICE IN AD HOC WIRELESS NETWORKS

AMITABH MISHRA
Johns Hopkins University

CAMBRIDGE
UNIVERSITY PRESS

CAMBRIDGE
UNIVERSITY PRESS

University Printing House, Cambridge CB2 8BS, United Kingdom

One Liberty Plaza, 20th Floor, New York, NY 10006, USA

477 Williamstown Road, Port Melbourne, VIC 3207, Australia

314-321, 3rd Floor, Plot 3, Splendor Forum, Jasola District Centre, New Delhi - 110025, India

103 Penang Road, #05-06/07, Visioncrest Commercial, Singapore 238467

Cambridge University Press is part of the University of Cambridge.

It furthers the University's mission by disseminating knowledge in the pursuit of education, learning and research at the highest international levels of excellence.

www.cambridge.org
Information on this title: www.cambridge.org/9780521878241

© Cambridge University Press 2008

First published 2008

A catalogue record for this publication is available from the British Library

ISBN 978-0-521-87824-1 Hardback

To my parents:
Shrimati Deomani and Shri Brij Mohan Lal Mishra

Contents

Preface

Security and quality of service in ad hoc wireless networks have recently become very important and actively researched topics because of a growing demand to support live streaming audio and video in civilian as well as military applications. While a couple of books have appeared recently that deal with ad hoc networks, a comprehensive book that deals with security and QoS has not yet appeared. I am confident that this book will fill that void.

The book grew out of a need to provide reading material in the form of book chapters to graduate students taking an advanced wireless networking course that I was teaching at the Virginia Polytechnic Institute and State University. Some of these book chapters then subsequently appeared as chapters in handbooks and survey papers in journals.

This book contains eight chapters in total, of which five chapters deal with various aspects of security for wireless networks. I have devoted only one chapter to the quality of service issue. Chapter 1 introduces basic concepts related to an ad hoc network, sets the scene for the entire book by discussing the vulnerabilities such networks face, and then produces a set of security requirements that these networks need to satisfy to live up to the challenges imposed by the vulnerabilities. Chapter 1 also introduces basic concepts regarding quality of service as it relates to ad hoc networks. In my presentation in this book, I have assumed that the reader is familiar with basic computer security mechanisms as well as the well known routing protocols of ad hoc networks.

Chapter 2 presents an overview of the wireless security for infrastructure-based wireless LANs that are based on the IEEE 802.11b standard, wireless cellular networks such as GSM, GPRS, and UMTS, and wireless personal area networks such as Bluetooth and IEEE 802.15.4 standard-based networks.

Various possible threats and attacks on ad hoc networks are discussed in Chapter 3. Possible security solutions against such attacks are then presented in various chapters of the book.

The security schemes that govern trust among communicating entities are collectively known as trust management. Chapter 4 presents various trust management schemes that are based on the principles and practice of key management in distributed networks and authentication. Chapter 5 addresses the issue of intrusion detection in ad hoc networks. It includes a discussion on both types of intrusion detection schemes, namely *anomaly* and *misuse* detection, and presents most of the prominent intrusion detection schemes available in the literature.

The topic of quality of service for ad hoc networks is covered in Chapter 6. Supporting appropriate quality of service for mobile ad hoc networks is a complex and difficult issue because of the dynamic nature of the network topology, and generally imprecise network state information. This chapter presents the basic concepts of quality of service support in ad hoc networks for unicast communication, reviews the major areas of current research and results, and addresses some new issues. Secure routing is the theme for Chapter 7, in which I describe the various algorithms that have been proposed to make the ad hoc routing more secure.

The IEEE 802.16 is a new standard that deals with providing broadband wireless access to residential and business customers and is popularly known as WiMax. This standard has several provisions for ensuring the security of and privacy to applications running on WiMax-enabled networking infrastructure. I discuss the security and privacy features of this standard in Chapter 8.

Acknowledgements

Among the people whose contributions helped me complete this book are Dr. Satyabrata Chakrabarti of Bell Laboratories, who was my guru, and Ketan Nadkarni, who was my graduate student at Virginia Tech. I thank both of them. I would also like to thank Dr. Philip Meyler, Editorial Manager at Cambridge University Press, for persuading me to complete this book. Without his support this book might not have been written at all. The entire Cambridge University Press team, including Anne Littlewood (Assistant Editor), Alison Lees (Copy-editor), and Daniel Dunlavey (Production Editor), has done an outstanding job in shaping this book to the final form, for which I am grateful.

Finally, I would like to thank my wife, Tanuja, and our children, Meghana and Anant, for making this book happen.

1

Introduction

Wireless mobile ad hoc networks consist of mobile nodes interconnected by wireless multi-hop communication paths. Unlike conventional wireless networks, ad hoc networks have no fixed network infrastructure or administrative support. The topology of such networks changes dynamically as mobile nodes join or depart the network or radio links between nodes become unusable. In this chapter, I will introduce wireless ad hoc networks, and discuss their inherent vulnerable nature. Considering the inherent vulnerable nature of ad hoc networks, a set of security requirements is subsequently presented. The chapter also introduces the quality of service issues that are relevant for ad hoc networks.

1.1 Ad hoc networking

Conventional wireless networks require as prerequisites a fixed network infrastructure with centralized administration for their operation. In contrast, so-called (wireless) mobile ad hoc networks, consisting of a collection of wireless nodes, all of which may be mobile, dynamically create a wireless network amongst themselves without using any such infrastructure or administrative support [1,2]. Ad hoc wireless networks are self-creating, self-organizing, and self-administering. They come into being solely by interactions among their constituent wireless mobile nodes, and it is only such interactions that are used to provide the necessary control and administration functions supporting such networks.

Mobile ad hoc networks offer unique benefits and versatility for certain environments and certain applications. Since no fixed infrastructure, including base stations, is prerequisite, they can be created and used "any time, anywhere." Such networks could be intrinsically fault-resilient, for they do not operate under the limitations of a fixed topology. Indeed, since all nodes are allowed to be mobile, the composition of such networks is necessarily time varying. Addition and deletion of nodes occur only by interactions with other

1

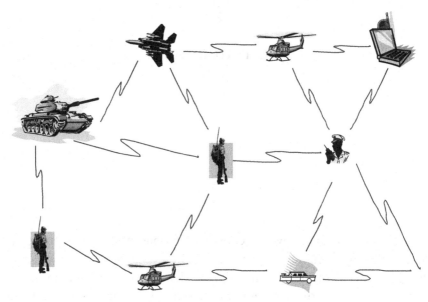

Figure 1.1 Conceptual representation of a mobile ad hoc network

nodes; no other agency is involved. Such perceived advantages elicited immediate interest in the early days among military, police, and rescue agencies in the use of such networks, especially under disorganized or hostile environments, including isolated scenes of natural disaster and armed conflict. See Fig. 1.1 for a conceptual representation. In recent days, home or small-office networking and collaborative computing with laptop computers in a small area (e.g., a conference or classroom, single building, convention center, etc.) have emerged as other major areas of application. These include commercial applications based on progressively developing standards such as Bluetooth [3], as well as other frameworks such as Piconet [4], HomeRF Shared Wireless Access Protocol [5], etc. In addition, people have recognized from the beginning that ad hoc networking has obvious potential use in all the traditional areas of interest for mobile computing.

Mobile ad hoc networks are increasingly being considered for complex multimedia applications, where various quality of service (QoS) attributes for these applications must be satisfied as a set of predetermined service requirements. As a minimum, the QoS issues pertaining to delay and bandwidth management are of paramount interest. In addition, because of the use of the ad hoc networks for military or police use, and of increasingly common commercial applications, various security issues need to be addressed. Cost-effective resolution of these issues at appropriate levels is essential for widespread general use of ad hoc networking.

Mobile ad hoc networking emerged from studies on extending traditional Internet services to the wireless mobile environment. All current works, as well as this presentation, consider the ad hoc networks as a wireless extension to the Internet, based on the ubiquitous IP networking mechanisms and protocols. Today's Internet possesses an essentially static infrastructure where network elements are interconnected over traditional wire-line technology, and these elements, especially the elements providing the routing or switching functions, do not move. In a mobile ad hoc network, by definition, all the network elements move. As a result, numerous more stringent challenges must be overcome to realize the practical benefits of ad hoc networking. These include effective routing, medium (or channel) access, mobility management, power management, and security issues, all of which affect the quality of the service experienced by the user.

The absence of a fixed infrastructure for ad hoc networks means that the nodes communicate directly with one another in a peer-to-peer fashion. The mobility of these nodes imposes limitations on their power capacity, and hence, on their transmission range; indeed, these nodes must often satisfy stringent weight limitations for portability. Mobile hosts are no longer just end systems; to relay packets generated by other nodes, each node must be able to function as a router as well. As the nodes move in and out of range with respect to other nodes, including those that are operating as routers, the resulting topology changes must somehow be communicated to all other nodes, as appropriate. In accommodating the communication needs of the user applications, the limited bandwidth of wireless channels and their generally hostile transmission characteristics impose additional constraints on how much administrative and control information may be exchanged, and how often. Ensuring effective routing is one of the great challenges for ad hoc networking.

The lack of fixed base stations in ad hoc networks means that there is no dedicated agency for managing the channel resources for the network nodes. Instead, carefully designed distributed medium access techniques must be used for channel resources, and, hence, mechanisms must be available to recover efficiently from the inevitable packet collisions. Traditional carrier sensing techniques cannot be used, and the hidden terminal problem [6,7] may significantly diminish the transmission efficiency [8]. An effectively designed protocol for medium access control (MAC) is essential to the quest for QoS.

1.2 The ad hoc wireless network: operating principles

I start with a description of the basic operating principles of a mobile ad hoc network. Figure 1.2 depicts the peer-level multi-hop representation of such a

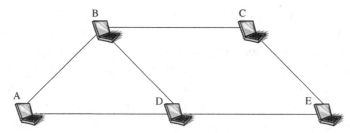

Figure 1.2 Example of an ad hoc network

network. Mobile node A communicates with another such node B directly (single-hop) whenever a radio channel with adequate propagation character-istics is available between them. Otherwise, multi-hop communication is necessary where one or more intermediate nodes must act as a relay (router) between the communicating nodes. For example, there is no direct radio channel (shown by the lines) between A and C or A and E in Fig. 1.2. Nodes B and D must, therefore, serve as intermediate routers for communication between A and C, and A and E, respectively. Indeed, a distinguishing feature of ad hoc networks is that all nodes must be able to function as routers on demand. To prevent packets from traversing infinitely long paths, an obvious essential requirement for choosing a path is that the path must be loop-free. A loop-free path between a pair of nodes is called a route.

An ad hoc network begins with at least two nodes broadcasting their presence (beaconing) with their respective address information. As discussed later, they may also include their location information, obtained, for example, by using a system such as the Global Positioning System (GPS), for more effective routing. If node A is able to establish direct communication with node B in Fig. 1.2, verified by exchanging suitable control messages between them, they both update their routing tables. When a third node, C, joins the network with its beacon signal, two scenarios are possible. The first is where both A and B determine that single-hop communication with C is feasible. In the second scenario, only one of the nodes, say B, recognizes the beacon signal from C and establishes the availability of direct communication with C. The distinct topology updates, consisting of both address and route updates, are made in all three nodes immediately afterwards. In the first case, all routes are direct. For the other, shown in Fig. 1.3, the route update first happens between B and C, then between B and A, and then again between B and C, confirming the mutual reachability between A and C via B.

The mobility of nodes may cause the reachability relations to change in time, requiring route updates. Assume that for some reason, the link between B and

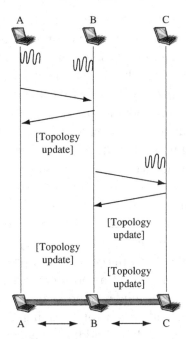

Figure 1.3 Bringing up an ad hoc network

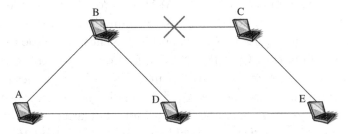

Figure 1.4 Topology update owing to a link failure

C is no longer available, as shown in Fig. 1.4. Nodes A and C can still reach each other, although this time only via nodes D and E. Equivalently, the original loop-free route $\langle A \leftrightarrow B \leftrightarrow C \rangle$ is now replaced by the new loop-free route $\langle A \leftrightarrow D \leftrightarrow E \leftrightarrow C \rangle$. All five nodes in the network are required to update their routing tables appropriately to reflect this topology change, which will be first detected by nodes B and C, then communicated to A and E, and then to D.

The reachability relation among the nodes may also change for other reasons. For example, a node may wander too far out of range, its battery may be depleted, or it may suffer a software or hardware failure. As more nodes join the network or some of the existing nodes leave, the topology updates become more numerous, complex, and, usually, more frequent, thus diminishing the network resources available for exchanging user information.

Finding a loop-free path as a legitimate route between a source–destination pair may become impossible if the changes in network topology occur too frequently. Here, "too frequently" means that there was not enough time to propagate to all the pertinent nodes all the topology updates arising from the last network topology changes, or worse, before the completion of determining all loop-free paths accommodating the last topology changes. The ability to communicate degrades with accelerating rapidity as the knowledge of the network topology becomes increasingly inconsistent. Given a specific time-window, we call (the behavior of) an ad hoc network combinatorially stable if, and only if, the topology changes occur sufficiently slowly to allow successful propagation of all topology updates as necessary. Clearly, combinatorial stability is determined not only by the connectivity properties of the networks, but also by the complexity of the routing protocol in use and the instantaneous computational capacity of the nodes, among other factors. Combinatorial stability is an essential consideration for attaining QoS objectives in an ad hoc network, as we shall see below. I address the general issue of routing in mobile ad hoc networks separately in the next section.

The shared wireless environment of mobile ad hoc networks requires the use of appropriate medium access control (MAC) protocols to mitigate the medium contention issues, allow efficient use of limited bandwidth, and resolve so-called hidden and exposed terminal problems. These are basic issues, independent of the support of QoS; the QoS requirements add extra complexities for the MAC protocols, mentioned later in Chapter 5. The issues of efficient use of bandwidth and the hidden/exposed terminal problem have been studied exhaustively and are well understood in the context of accessing and using any shared medium. I briefly discuss the "hidden-terminal" problem [6] as an issue especially pertinent for the wireless networks.

Consider the scenario of Fig. 1.5, where a barrier prevents node B from receiving the transmission from D, and vice versa, or, as usually stated, B and D cannot "hear" each other. The "barrier" does not have to be physical; a large enough distance separating two nodes is the most commonly occurring "barrier" in ad hoc networks. Node C can "hear" both B and D. When B is transmitting to C, D, being unable to "hear" B, may transmit to C as well, thus causing a collision and exposing the *hidden-terminal* problem. In this case, B and D are "hidden" from each other. Now consider the case when C is transmitting to D. Since B can "hear" C, B cannot risk initiating a transmission to A for fear of causing a collision at C. Here is an example of the *exposed terminal* problem, where B is "exposed" to C.

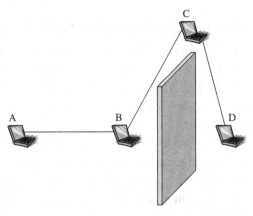

Figure 1.5 Example of hidden/exposed terminal problem

A simple message exchange protocol solves both problems. When D wishes to transmit to C, it first sends a request-to-send (RTS) message to C. In response, C broadcasts a clear-to-send (CTS) message that is received by both B and D. Since B has received the CTS message unsolicited, B knows that C is granting permission to send to a hidden terminal and hence refrains from transmitting. Upon receiving the CTS message from C in response to its RTS message, D transmits its own message.

Not only does the above (crude and deliberately simplified outline of the) dialogue solve the hidden terminal problem, but it solves the exposed terminal problem as well, for after receiving an unsolicited CTS message, B refrains from transmitting and cannot cause a collision at C. After an appropriate interval, determined by the attributes of the channel (i.e., duration of a time slot, etc.), B can send its own RTS message to C as the prelude to a message transmission.

Limitation on the battery power of the mobile nodes is another basic issue for ad hoc networking. Limited battery power restricts the transmission range (hence the need for each node to act as a router) as well as the duration of the active period for the nodes. Below some critical thresholds for battery power, a node will not be able to function as a router, thus immediately affecting the network connectivity, possibly isolating one or more segments of the network. Fewer routers almost always mean fewer routes and, therefore, increased likelihood of degraded performance in the network. Indeed, QoS obviously becomes meaningless if a node is not even able to communicate, owing to low battery power. Since exchange of messages necessarily means power consumption, many ad hoc networking mechanisms, especially routing and security protocols, explicitly include minimal battery power consumption as a design objective.

1.3 Ad hoc networks: vulnerabilities

There are various reasons why wireless ad hoc networks are at risk, from a security point of view. I next discuss the characteristics that make these networks vulnerable to attacks. Attacks are procedures that are launched by unauthorized entities or nodes within the networks to disrupt the normal operation of the enterprise.

The wireless links between nodes are highly susceptible to link attacks, which include passive eavesdropping, active interfering, leaking secret information, data tampering, impersonation, message replay, message distortion, and denial of service. Eavesdropping might give an adversary access to secret information, violating confidentiality. Active attacks might allow the adversary to delete messages, to inject erroneous messages, to modify messages, and to impersonate a node, thus violating availability, integrity, authentication, and non-repudiation (these and other security needs are discussed in the next section).

Ad hoc networks do not have a centralized piece of machinery such as a name server or a base station, which could lead to a single point of failure and, thus, make the network that much more vulnerable. On the flipside, however, the lack of support infrastructure leads to prevention of application of standard techniques such as key management (discussed later in the book) to secure the network. This gives rise to the need for new schemes to ensure key agreement.

An additional problem that arises in ad hoc networks is the accurate detection of a compromised node. Usually compromised nodes are detected by monitoring their behavior. But in a wireless environment it is often difficult to distinguish between a truly misbehaving node and a node that appears to be misbehaving because of poor link quality. The presence of compromised nodes has the potential to cause Byzantine failures, which are encountered within mobile ad hoc network (MANET) routing protocols, wherein a set of the nodes could be compromised in such a way that the incorrect and malicious behavior cannot be directly noted at all. The compromised nodes may seemingly operate correctly, but, at the same time, they may make use of the flaws and inconsistencies in the routing protocol to distort the routing fabric of the network. In addition, such malicious nodes can also create new routing messages and advertize non-existent links, provide incorrect link state information and flood other nodes with routing traffic, thus inflicting Byzantine failures on the system. Such failures are especially severe because they may come from seemingly trusted nodes, whose malicious intentions have not yet been noted. Even if the compromised nodes were noticed and prevented from performing incorrect actions, the erroneous information generated by the Byzantine failures could have already been propagated through the network.

No part of the network is dedicated to support any specific network functionality. All nodes are expected to contribute to routing (topology discovery, data forwarding). The examples of functions that rely on a central service, and which are also of high relevance, are naming services, certification authorities, directory, and other administrative services. In ad hoc networks, nodes cannot rely on such a service. Even if such services were assumed, their availability would not be guaranteed, either due to the dynamically changing topology that could easily result in a partitioned network, or due to congested links close to the node acting as a server.

The absence of infrastructure and the consequent absence of authorization facilities impede the usual practice of establishing a line of defence, distinguishing nodes as trusted and non-trusted. Such a distinction would have been based on a security policy, the possession of the necessary credentials and the ability of nodes to validate them. In the case of wireless ad hoc networks, there may be no grounds for such a priori node classification, since all nodes are required to cooperate in supporting the network operation, while no prior security association can be assumed for all the network nodes.

Additionally, freely roaming nodes form transient associations with their neighbors; they join and leave sub-domains independently and without notice. Thus, it may be difficult, in most cases, to have a clear picture of the ad hoc network membership at a given time. Consequently, especially in the case of a large network, no form of established trust relationships among the majority of nodes can be assumed.

In such an environment, there is no guarantee that a path between two nodes would be free of malicious nodes. There is a possibility that a path consisting of malicious nodes may not comply with the rules of the protocol employed and can attempt to disrupt the network operation. The mechanisms currently incorporated in ad hoc routing protocols cannot cope with disruptions due to malicious behavior. For example, any node could claim that it is one hop away from the sought destination, causing all routes to the destination to pass through itself. Alternatively, a malicious node could corrupt any in-transit route request (reply) packet and cause data to be misrouted.

The presence of even a small number of adversarial nodes could result in repeatedly compromised routes, and, as a result, the network nodes would have to rely on cycles of timeout and new route discoveries to communicate. This would incur arbitrary delays before the establishment of a non-corrupted path, while successive broadcasts of route requests would impose excessive transmission overhead. In particular, intentionally falsified routing messages would result in a denial-of-service (DoS) experienced by the end nodes.

The dynamic and transient nature of an ad hoc network can result in constant changes in trust among nodes. This can create problems, for example, with key management, if cryptography is used in the routing protocol. It must not be trivial, for example, to recover private keys from the device. Evidence that tampering has occurred would be required so as to distinguish a tampered node from the rest. Standard security solutions would not be good enough since they are essentially for statically configured systems. This gives rise to the need for security solutions, which adapt to the dynamically changing topology and movement of nodes in and out of the network.

Moreover, the battery-powered operation of ad hoc networks gives attackers ample opportunity to launch a denial-of-service attack by creating additional transmissions or expensive computations to be carried out by a node in an attempt to exhaust its batteries.

In addition, sensor networks (a form of wireless ad hoc network) are made up of devices that tend to have limited computational abilities. For example, the working memory of a sensor node is insufficient even to hold the variables (of sufficient length to ensure security) that are required in asymmetric cryptographic algorithms, let alone perform operations on them. This may exclude techniques such as frequent public key cryptography during normal operation. A particular challenge is that of broadcasting authenticated data to the entire sensor network. Current proposals for authenticated broadcast rely on asymmetric digital signatures for the authentication, and these are impractical for many reasons (e.g., long signatures with high communication overheads of 50–1000 bytes per packet; very high overheads to create and verify the signature) for sensor networks.

Lastly, scalability is another issue, which has to be addressed when security solutions are being thought of, for the simple reason that an ad hoc network may consist of hundreds or even thousands of nodes. Many ad hoc networking protocols are applied in conditions where the topology must scale up and down efficiently, e.g., because of network partitions or mergers. The scalability requirements here refer to the scalability of individual security services such as key management for example.

The above discussion makes it clear that ad hoc networks are inherently insecure, more so than their wireline counterparts, and need robust security schemes that take into consideration the inherently susceptible nature of these networks. Coming up with a security scheme, in general, necessitates the discussion of the fundamental components that make up security. In the next section, I take a look at the essential security needs of such networks. By this, I mean the factors that ought to be taken into consideration when designing a security scheme.

1.4 Ad hoc networks: security requirements

Security is a term that is liberally used in computer networks terminology. In this section I will go over the several attributes and terms that define security and are often used in security-related discussions, in the context of computer networks. The basic security needs of wireless ad hoc networks are more or less the same as those of wired networks. To some extent, several security schemes of the wire-line networks have been developed and implemented in wireless cellular networks. To make ad hoc networks secure, we need to find ways to incorporate some of these schemes of wireless and wire-line networks. I devote several chapters of this book to address incorporation of these schemes in ad hoc networks. In the following, I briefly introduce the standard terms, which are used when security aspects of a network are discussed.

(1) *Availability*
    The services provided by a node continue to be provided irrespective of attacks. Nodes should be available for communication at all times. In other words, availability ensures survivability of the network services in presence of denial-of-service (DoS) attacks, which can be launched at any layer of an ad hoc network through radio jamming or battery exhaustion.

(2) *Authenticity*
    This is essentially a confirmation that parties, in communication with each other, are genuine and not impersonators. This would require the nodes to somehow prove that their identities are what they claim to be. Without authentication, an adversary could very well masquerade a node, could get access to sensitive and classified information, and could even interfere with the normal and secure network operation.

(3) *Confidentiality*
    This ensures that information is not disclosed to unauthorized entities, i.e., an outsider should not be able to access information in transit between two nodes. Confidentiality necessitates the prevention of intermediate and non-trusted nodes from understanding the content of the packets being transmitted. If authentication is taken care of properly, then confidentiality is a relatively simple process.

(4) *Integrity*
    This is the guarantee that the message or packet being delivered has not been modified in transit or otherwise, and that what has been received is what was originally sent. A message could be corrupted owing to non-malicious reasons, such as radio propagation impairment, but there is always the possibility that an adversary has maliciously modified the content of the message.

(5) *Non-repudiation*
    The sender of a message cannot later deny sending the information or the receiver cannot deny the reception. This can come in handy while detecting and isolating compromised nodes. Any node, which receives an erroneous message, can accuse

the sender with proof and thus, convince other nodes about the compromised node. Routers cannot repudiate ownership of routing protocol messages they send. The trust associated with the propagation of updates that originate from distant nodes forms a major concern.

(6) *Ordering*

Updates received from routers are in order, the non-occurrence of which can affect the correctness of routing protocols. Messages may not reflect the true state of the network and may propagate false information.

(7) *Timeliness*

Routing updates should be delivered in a timely fashion. Update messages that arrive late may not reflect the true state of links or routers on the network. They can cause incorrect forwarding or even propagate false information and weaken the credibility of the update information. If a node that relays information between two highly connected components is advertised as "down" by malicious neighbors, a large part of the network becomes unreachable.

(8) *Isolation*

This requires that the protocol be able to identify misbehaving nodes and make them unable to interfere with routing. Alternatively, the routing protocol should be designed to be immune to malicious nodes.

(9) *Authorization*

An authenticated user or node is issued an unforgeable credential by the certificate authority. These credentials specify the privileges and permissions associated by the users or the nodes. Currently, credentials are not used in routing protocol packets, and any packet can trigger update propagations and modifications to the routing table.

(10) *Lightweight computations*

Many devices connected to an ad hoc network are assumed to be battery-powered with limited computational abilities. Such a node cannot be expected to be able to carry out expensive computations. If operations such as public key cryptography or shortest path algorithms for large networks prove necessary, they should be confined to the least possible number of nodes; preferably only the route end points at route creation time.

(11) *Location privacy*

Often, the information carried in message headers is just as valuable as the message itself. The routing protocol should protect information about the location of nodes in a network and the network structure.

(12) *Self-stabilization*

A routing protocol should be able to recover automatically from any problem in a finite amount of time without human intervention. That is, it must not be possible to permanently disable a network by injecting a small number of malicious packets. If the routing protocol is self-stabilizing, an attacker who wishes to inflict continuous damage must remain in the network and continue sending malicious data to the nodes, which makes the attacker easier to locate.

(13) *Byzantine robustness*

A routing protocol should be able to function correctly even if some of the nodes participating in routing are intentionally disrupting its operation. Byzantine robustness can be seen as a stricter version of the self-stabilization property: the routing protocol must not only automatically recover from an attack; it should not cease from functioning even during the attack. Clearly, if a routing protocol does not have the self-stabilization property, it cannot have Byzantine robustness either.

(14) *Anonymity*

Neither the mobile node nor its system software should expose any information that allows any conclusions about the owner or current user of the node. In case device or network identifiers are used (e.g., MAC address, IP address), no linking should be possible between the respective identifier and the owner's identity for the communication partner or any outside attacker.

(15) *Key management*

The services in key management must provide solutions to the following questions:

- *Trust model* – how many different elements in the network can trust each other and trust relationships between network elements;
- *Cryptosystems* – while public-key cryptography offers more convenience, public-key cryptosystems are significantly slower than their secret-key counterparts when a similar level of security is needed;
- *Key creation* – which parties are allowed to generate keys to themselves or other parties, and what kind of keys;
- *Key storage* – any network element may have to store its own key and possibly keys of other elements as well, while in systems with shared keys with parts of keys distributed to several nodes, the compromising of a single node does not yet compromise the secret keys;
- *Key distribution* – generated keys have to be securely distributed to their owners, and any key that must be kept secret has to be distributed so that confidentiality, authenticity, and integrity are not violated.

(16) *Access control*

This consists of the means to govern the way the users or virtual users such as operating system processes (subjects) can have access to data (objects). Only authorized nodes may form, destroy, join, or leave groups. Access control can also mean the way the nodes log into the networking system to communicate with other nodes when initially entering the network. There are various approaches to access control: discretionary access control (DAC) offers means for defining the access control to the users themselves; mandatory access control (MAC) involves centralized mechanisms to control the access to objects with formal authorization policy. Finally, role based access control (RBAC) applies the concept of roles within the subjects and objects.

(17) *Trust*

If physical security is low and trust relationships are dynamic, then the probability of a security failure may rise rapidly. It is not difficult to see what happens if the suspicion of a security failure increases. If there is a reason to believe that a part of the nodes belonging to a network have been compromised, users will probably become more reluctant to trust the network. Constructing security for the first time may not be so difficult. Maintaining trust and handling dynamic changes over time seem to need more effort.

In summary, we can safely say that the mandatory security requirements include confidentiality, authentication, integrity, and non-repudiation. These would, in turn, require some form of cryptography, certificates, and signatures. Some other ideal characteristics include user authentication, explicit transaction authorization, end-to-end encryption, accepted log-on security (biometrics) instead of separate personal identification numbers (PINs) and passwords, intrusion detection, access control, logging, audit trail, security policy that states the rules for access, anti-virus scanners for the content, firewall, etc. This discussion demarcates the various branches within security, per se, such as intrusion detection and prevention, key agreement, trust management, data encryption, and access control. Having looked at the essential security needs, we are now ready to discuss the various kinds of attacks, practical as well as conceptual. This discussion forms the basis of Chapter 3.

Having discussed basics of the security needs for ad hoc networks, I now introduce the challenges associated with providing quality of service (QoS) in ad hoc networks. It should be pointed out that security and quality of service are two distinct attributes that are independent of each other in general. For example a secure routing protocol may have no QoS features in it or a QoS-based routing algorithm may not be secure. There can be some dependence on each other: if both features are part of the network architecture, then one can have an impact on the other. For example, a heavy computational burden imposed by a cryptography algorithm may affect the delay at one of the nodes. Our treatment in this book is confined to treating the security and QoS aspects related to ad hoc networks as independent.

1.5 Quality of service

All the vulnerabilities enumerated in Section 1.3 above are potential sources of service impairment in ad hoc networks and hence may degrade the "quality of service" seen by the users. As of now, the Internet has only supported "best effort" service – best effort in the sense that it will do its best to transport the user packets to their intended destination, although without any guarantee. Quality of service support is recognized as a challenging issue for the Internet, and a vast amount of research on this issue has appeared in the literature during the last decade or so [9]. With the Internet as the basic model, ad hoc networks have been initially considered only for "best effort" services as well, especially given their peculiar challenges when compared against traditional wire-line or even conventional wireless networks. Indeed, just as the QoS accomplishments for wired networks such as the Internet cannot be directly extended to the wireless environment, the QoS issues become even more formidable for mobile ad hoc networks. Happily, during the last few years, QoS for ad hoc networks has emerged as an active and fertile research topic of a growing number of researchers and many major advances are expected in the next few years.

Performance of these various protocols under "field" conditions is, of course, the final determinant of their efficacy and applicability. Relative comparisons of computational and communication complexities of various routing protocols for ad hoc networks have appeared in the past, providing the foundation for more application-oriented assessment of their effectiveness. On the other hand, the performance studies have started to appear only recently. The mathematical analysis of ad hoc networks, even under the simplest assumptions about the dynamics of topology changes and traffic processes, poses formidable challenges, and even their simulation is considerably more difficult than their static counterparts. Performance studies of ad hoc networks with QoS constraints continue to be an active area of research. Chapter 6 discusses the state of the art of quality of service in ad hoc networks and is a good source of more up-to-date information in this area.

1.6 Further reading

This chapter introduced the basic concepts of ad hoc networks and exposed their inherent vulnerable nature. To address their vulnerabilities, several security requirements have been proposed in the literature, which are also presented. As these networks are maturing, interest has been growing in supporting real-time traffic on ad hoc networks. Support of real-time traffic on a packet network requires that the network is able to meet stringent quality of service requirements such as delay and jitter, which are briefly discussed. To get a better understanding of ad hoc networking concepts, I recommend reading any of the following fine books: [10,11,12, and 13].

1.7 References

[1] Z. J. Haas, M. Gerla, D. B. Johnson, *et al.*, "Guest editorial," *IEEE J. Select. Areas Commun.*, Special issue on wireless networks, vol. 17, no. 8, Aug. 1999, pp. 1329–1332.

[2] D. B. Johnson and D. A. Maltz, "Protocols for adaptive wireless and mobile networking," *IEEE Personal Commun.*, Feb. 1996, pp. 34–42.

[3] C. Bisdikian, "An overview of the Bluetooth wireless technology," *IEEE Commun. Mag.*, Dec. 2001, pp. 86–94. (For additional sources of comprehensive information on Bluetooth, see the official websites, www.bluetooth.com/ and www.bluetooth.org/; an excellent compendium of tutorials and references is available at http://kjhole.com/Standards/Intro.html.)

[4] F. Bennett, D. Clarke, J. B. Evans, *et al.*, "Piconet: embedded mobile networking," *IEEE Personal Commun.*, vol. 4, no. 5, Oct. 1997, pp. 8–15.

[5] K. J. Negus, J. Waters, J. Tourrilhes, *et al.*, "HomeRF and SWAP: wireless networking for the connected home," *ACM SIGMOBILE Mobile Computing and Commun. Rev.*, vol. 2, no. 4, Oct. 1998, pp. 28–37.

[6] F. A. Tobagi and L. Kleinrock, "Packet switching in radio channels - part 2: the hidden terminal problem in carrier sense multiple-access and the busy tone solution," *IEEE Trans. Commun.*, vol. COM-23, Dec. 1985, pp. 1417–1433.

[7] C. R. Lin and M. Gerla, "MACA/PR: an asynchronous multimedia multihop wireless network," *Proc. 16th Annual Joint Conf. IEEE Comp. Commun. Soc. (INFOCOM 1997)*, vol. 1, 1997, pp. 118–125.

[8] J. L. Sobrinho and A. S. Krishnakumar, "Quality-of-service in ad hoc carrier sense multiple access wireless networks," *IEEE J. Select. Areas Commun.*, vol. 17, No. 8, Aug. 1999, pp. 1353–1414.

[9] S. Chen and K. Nahrstedt, "An overview of quality-of-service routing for the next generation high-speed networks: problems and solutions," *IEEE Network*, Nov.–Dec. 1998, pp. 64–79.

[10] S. Basagni, M. Conti, S. Giordano, and I. Stojmenovic (Editors), *Mobile Ad Hoc Networking*, John Wiley and Sons, 2004.

[11] M. Ilyas (Editor), *The Handbook of Wireless Ad Hoc Networks*, CRC Press, 2003.

[12] C. S. Ram Murthy and B. S. Manoj, *Ad Hoc Wireless Networks – Architecture and Protocols*, Prentice Hall, 2004.

[13] I. Stojmenovic (Editor), *Handbook of Wireless Networks & Mobile Computing*, John Wiley and Sons, 2002.

2

Wireless security

Wireless networks are typically divided into three classes depending on their range of transmissions. We have personal area networks (PANS) that have a very low transmission range, of the order of several meters; Bluetooth happens to be the representative network or technology when wireless personal area networks are mentioned. On a slightly larger transmission scale, of the order of 100–200 meters, we have wireless local area networks (LANs), known as 802.11 or WiFi, which are very well deployed all over the world. The personal area and local area networks have been primarily designed for indoor applications. Networks that have transmission in the range of several kilometers are known as wireless wide area networks (WANs), and cellular networks of different vintages are prime examples of such networks. So any discussion of security in a wireless environment will not be complete unless the proposed security schemes for these three distinct networks are examined. In this chapter, I briefly go over the security schemes of wireless PAN, LAN, and WAN networks. For readers interested in knowing more about these topics, appropriate references are highlighted. I begin this chapter by discussing WiFi security, followed by cellular network security, and concluding with the security of personal area networks.

2.1 Wireless local area networks (IEEE 802.11) security

2.1.1 Introduction

A wireless local area network (WLAN) is a flexible data communication system implemented as an extension to, or as an alternative to, a wired LAN. Wireless local area networks transmit and receive data over the air via RF technology, minimizing the need for any wired connections, and in turn, combining data connectivity with user mobility. They provide all the functionalities of LANs

17

without the physical constraints, and their configurations vary from a simple peer-to-peer topology to complex networks offering distributed data connectivity and roaming.

The market for wireless communication has grown rapidly since the introduction of the IEEE 802.11b wireless local area networking standard, which offers performance more nearly comparable to that of an Ethernet. The 802.11b standard, published in September 1999 [1], can deliver data rates up to 11 Mbps. The 802.11b standard specifies the lowest layer of OSI network model (i.e., physical layer) and a part of the next higher layer (data link layer). In addition, the standard specifies the use of Ethernet protocol (IEEE 802.3) for the logical link control (LLC) portion of the data link layer. Higher layer protocols are TCP/IP and applications that can run on top of TCP/IP.

Wireless LAN devices are equipped with a special network interface card (NIC) with one or more antennae, a radio receiver, and circuitry to convert between the analog radio signals and the digital pulses used by the computers. Radio waves broadcast on a given frequency can be picked by any receiver within the range tuned to that frequency. Effective and usable range depends on signal power, distance, and interference from intervening objects or other signals. A typical range of a wireless transmission in 802.11b is in the hundreds of meters. The full set of data rates in this standard is 11, 5.5, 2, and 1 Mbps.

The 802.11 mobile station may be mobile, portable, or stationary. Mobile stations dynamically associate with wireless LAN cells, or *basic service sets* (BSSs). The 802.11 MAC protocol supports the formation of two distinct types of BSS. The first type is the independent BSS, or ad hoc BSS. Ad hoc BSSs are self-forming; they are created and maintained as needed without prior administrative arrangements, often for specific purposes (such as transferring a file from one personal computer to another). Stations in an ad hoc BSS establish MAC layer wireless links with those stations in the BSS with which they desire to communicate, and frames are transferred directly from source to destination stations. Therefore, stations in an ad hoc BSS must be within range of one another to communicate. Furthermore, no architectural provisions are made for connecting the ad hoc BSSs to external networks, so communication is limited to stations within the ad hoc BSS.

The second type of BSS is the infrastructure BSS; this is more commonly used in practice. This type supports extended interconnected wireless and wired networking. Within each infrastructure BSS is an access point (AP), a special central traffic relay station that normally operates on a fixed channel and is stationary. Access points connect the infrastructure BSS to an IEEE abstraction known as *distribution system* (DS). Multiple APs connected to a common DS form an extended service set (ESS). A distribution system is

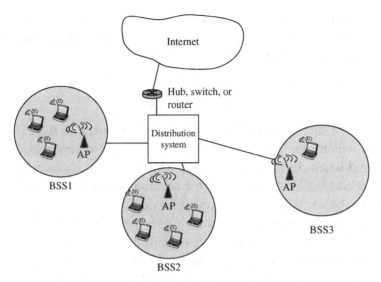

Figure 2.1 An 802.11 network with infrastructure

usually connected to a switch, a hub, or a router through which access to other networks, such as the Internet, is possible. The DS is responsible for forwarding frames within the ESS, between APs and the switch or the router, and it may be implemented with wired or wireless links. See Fig. 2.1.

Mobile stations in an infrastructure BSS establish MAC layer links with an AP. Furthermore, they only communicate directly to and from the selected AP. The AP/DS utilizes store and forward retransmission for intra-BSS traffic to provide connectivity between the mobile stations in the BSS. Typically, at most, only a small fraction of the frames flows between mobile stations within an infrastructure BSS; therefore retransmission results in a small overall bandwidth penalty. The effective physical span of BSS is of the order of twice the maximum mobile station-to-station range; mobile stations must be within range of the AP to join BSS but may not be within range of all other mobile stations in the BSS.

Mobile stations utilize 802.11 architected scan, authentication, and association processes to join an infrastructure BSS and connect to the wireless LAN system. Scanning allows mobile stations to discover existing BSSs that are within range. Access points periodically transmit beacon frames that, among other things, may be used by mobile stations to discover BSSs. Before joining a BSS, a mobile station must demonstrate through authentication that it has credentials to join. The actual BSS join occurs through association. Mobile stations can be authenticated by multiple APs but may be associated with only one AP at a time. Roaming mobile stations initiate handoff from one BSS to another through reassociation. The reassociation management frame is both a request by the sending mobile station to disassociate from the currently associated BSS and a request to join a new BSS.

2.1.2 Medium access

One of the most significant differences between Ethernet and 802.11b LANs is the way in which they control access to the medium, determining who may transmit and when. Ethernet uses carrier sense multiple access with collision detection (CSMA/CD). This is possible because an Ethernet device can send and listen to the wire signal at the same time, detecting patterns that show that a collision is taking place. When a radio attempts to transmit and listen on the same channel at the same time, its own transmission drowns out all other signals. Collision detection is impossible.

The carrier sense capabilities of Ethernet and wireless LANs are also different. On an Ethernet segment, all stations are within range of one another at all times, by definition. When the medium seems clear, it is clear. Only a simultaneous start of transmissions results in a collision. Nodes in a wireless LAN cannot always tell by listening alone whether or not the medium is, in fact, clear. In wireless LAN, it is possible to have hidden terminals (as described in Chapter 1); a situation that arises when two nodes hear a third node clearly but cannot hear each other.

To solve the hidden node problem and overcome the impossibility of collision detection, 802.11b wireless LANs use CSMA/CA (carrier sense multiple access with collision avoidance). Under CSMA/CA, devices use a four-way handshake (RTS/CTS/DATA/ACK) to gain access to the airwaves and ensure collision avoidance. Here RTS, CTS, DATA, ACK stand for request-to-send, clear-to-send, data, and acknowledgement. See [1] for four-way handshake and other timing-related waiting periods. To send a direct transmission to another node, the source node puts a short request-to-send (RTS) packet on the air, addressed to the intended destination. If that destination hears the transmission and is able to receive, it replies with a short clear-to-send (CTS) packet. The initiating node then sends the data, and the recipient acknowledges all transmitted packets by returning a short acknowledgement (ACK) packet for every transmitted packet received. The 802.11 standard also implements a truncated binary backoff, in case multiple nodes are trying to access the medium simultaneously. The 802.11b standard describes the backoff mechanism in detail.

Timing is critical to mediating access to the airwaves in wireless LANs. To ensure synchronization, access points or their functional equivalents periodically send beacons and timing information.

2.1.3 Authentication and privacy

Wireless LANs are subject to possible unwanted monitoring. For this reason, IEEE 802.11 specifies an optional MAC layer security system known as *wired equivalent privacy* (WEP). As the name implies, WEP is intended to provide to the wireless Ethernet a level of privacy similar to that enjoyed by wired Ethernets. Wired equivalent privacy involves a shared key authentication service with RC4 encryption. This is a stream cipher designed by Ronald Rivest of the RSA Security algorithm, and is commonly known as Ron's Cipher 4. Ron's Cipher 4 is used to generate a pseudo-random number sequence that is "XORed" into the data stream. A key, derived by combining a secret key and an initialization vector (IV), is used to set the initial condition or the state of the RC4 pseudo-random number generator. By default, each BSS supports up to four 40 bit keys that are shared by all the stations in the BSS. Keys unique to a pair of communicating stations and direction of transmission may also be used (that is, unique to a transmit–receive address pair). Key distribution is outside the scope of the standard but presumably utilizes a secure mechanism.

When a station attempts to authenticate with a second station that implements WEP, the authenticating station presents challenge text to the requesting station. The requesting station encrypts the challenge text using the RC4 algorithm and returns the encrypted text to the authenticating station. The encrypted challenge text is decrypted and checked by the authenticating station before completing authentication. After authentication and association, the frame body (the MAC payload) is encrypted in all frames exchanged between the stations. Encrypted frames are decrypted and checked by the MAC layer of receiving stations before being passed to the upper protocol layers.

Operation of WLANs is governed by the IEEE 802.11b standard, which defines two native mechanisms for providing access control and privacy on wireless LANs: service set identifiers (SSIDs) and wired equivalent privacy (WEP). Another mechanism to ensure privacy through encryption is by using the virtual private network (VPN) that runs transparently over a wireless LAN. In this section I discuss native schemes as well as non-native VPN based security schemes for IEEE 802.11 WLANs.

2.1.4 Native security schemes

Service set identifiers

One commonly used wireless local area network feature is a naming handle called "service set identifier" (SSID). This provides a rudimentary level of access control. An SSID is a common network name for the devices in a wireless local area network subsystem. The SSID serves to segment that subsystem logically. The use of the SSID as a handle to authorize system access can be dangerous because SSID is itself not well secured. An access point (AP) that connects wireless LAN to the wired LAN is usually set to broadcast its SSID in its beacons.

Wired equivalent privacy (WEP)

The IEEE 802.11b standard provides an optional encryption scheme called wired equivalent privacy (WEP) that offers a mechanism for securing wireless LAN data streams. Wired equivalent privacy is based on a symmetric key scheme, in which the same key and algorithms are used for both encryption and decryption of data. The objectives of WEP are:

(1) *Access control*: prevention of unauthorized access to the system without a correct WEP key;
(2) *Privacy*: protection of wireless LAN data streams by encrypting them and allowing decryption only for the users with the correct WEP keys.

Although WEP is optional, support for WEP with 40 bit encryption keys is a requirement for Wi-Fi certification by WECA (the Wireless Ethernet Compatibility Alliance), so WECA members generally support WEP. Wired equivalent privacy is implemented in software by some WLAN vendors while others implement it in hardware accelerators to minimize the performance degradation of encrypting and decrypting data streams.

The IEEE 802.11 standard provides two schemes for defining WEP keys to be used on WLANs. With the first scheme, a set of as many as four default keys is shared by all stations (i.e., clients and access points) in a wireless subsystem. When a client obtains the default keys, that client can communicate securely with all other stations in the subsystem. The problem with the default keys is that when they become widely distributed they are more likely to be compromised. In the second scheme, each client establishes a key mapping relationship with another station: this is a more secure operation because fewer stations have the keys. The distribution of unicast keys becomes more difficult as the number of stations increases.

Authentication

A user cannot participate in a wireless LAN until that client is authenticated. The IEEE 802.11b standard defines two types of authentication methods: open and shared key. The authentication method must be set on each client and the setting should match that of the access point with which the client wants to associate. With open authentication, which is the default, the entire authentication process is handled in the clear text, and a client can associate with an access point even without supplying the correct WEP key. With the shared key authentication, the access point sends the client a challenge packet that the client must encrypt with the correct WEP key and return to the access point. If the client has the wrong key or no key, it will fail authentication and will not be allowed to associate with the access point.

Some LAN vendors support authentication based on the physical address, or medium access control (MAC) address of a client. An access point will allow association by a client only if that client MAC address matches an address in an authentication table used by the access point.

2.1.5 Security threats

Wireless LANs are exposed to several security threats, and so require protection against such threats. In the following, I discuss common threats and possible solutions.

Stolen hardware

Generally, it is common to assign, statically, a WEP key to the client, either on the client's disk storage or in the memory of the client's wireless LAN adaptor. When this is done, the possessor of a client has the possession of the client's MAC address and WEP key and can use those components to gain access to the wireless LAN. If multiple users share a client, then those users effectively share the MAC address and WEP key. When a client is lost or stolen, the intended user or users of the client no longer have access to the MAC address or WEP key and an unintended user does. It is almost impossible for an administrator to detect the security breach; a legitimate owner must inform the administrator, who in turn will render the MAC address and WEP key useless for wireless LAN access and decryption of transmitted data. The administrator must recode static encryption keys on all clients that use the same keys as the lost or stolen client. The greater the number of clients, the bigger is the task of reprogramming the WEP keys. This situation calls for a security solution that:

(1) Has device independent authentication procedures such as those that use usernames and passwords, thereby allowing independence from the hardware;
(2) Has WEP keys that are dynamically generated after user authentication, instead of static keys that are associated with particular clients.

Malicious access points

The 802.11b shared key authentication procedure employs one-way authentication. For example, an access point authenticates a user, but a user does not and cannot authenticate an access point. If a malicious access point is placed on a wireless LAN, it can be a launch pad for denial of service attacks through the hijacking of legitimate users. What is needed is a mutual authentication between the client and an authentication server which allows the legitimacy of both sides to be proved within a reasonable time. Because a client and an authentication server communicate through an access point, the access point must support the mutual authentication scheme that allows for the detection and isolation of malicious access points.

Miscellaneous threats

The standard version of WEP supports per-packet encryption but not per-packet authentication and, as a result, is vulnerable to spoofing. One way to mitigate this security weakness is to ensure that WEP keys are changed frequently. By monitoring the 802.11 control and data channels, a hacker can obtain information such as:

(1) Client and access point MAC addresses;
(2) The MAC addresses of internal hosts;
(3) Times of association and disassociation.

The hacker may use some of this information for long-term traffic profiling and analysis that may provide user or device specific information. To mitigate such weaknesses, it is appropriate to use per-session WEP keys.

2.1.6 Dealing with security threats

Wireless LAN security concerns can be addressed by adopting schemes that:

(1) Use authentication procedures that are independent of devices. Examples are usage of usernames and passwords.
(2) Use mutual authentication between a client and an authentication RADIUS server.
(3) Use dynamically generated WEP keys for user authentication.
(4) Use session-based WEP keys.

Currently, there are two major approaches to deal with wireless LAN security issues. One approach that has been embraced by several vendors is based on using an extensible authentication protocol (EAP) with the IEEE 802.1× protocol, and the other is based on using a virtual private network. I discuss both of these approaches in the next two sections.

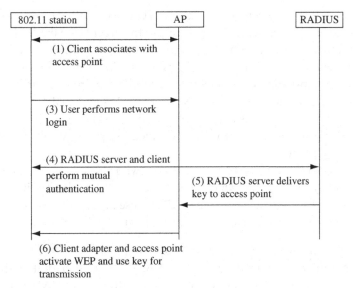

Figure 2.2 Message flows in 802.1× with EAP

Extensible authentication protocol with IEEE 802.1×

When a security solution that uses EAP and 802.1× is in place, a wireless client that associates with an access point cannot gain access to the network until the user performs a network logon. When the user provides a username and password to the network, the client and RADIUS server perform mutual authentication, with the client authenticated by the supplied username and password. The RADIUS server and client then derive a client-specific WEP key to be used by the client for the current logon session. All sensitive information, such as the password, is protected from passive monitoring and other methods of attack. Nothing is transmitted over the air that is clear.

Figure 2.2 illustrates the sequence of events that take place before a user starts to access the system services:

(1) A wireless client associates with an access point.
(2) The access point blocks all attempts by the client to gain access to network resources until the client logs on to the network.
(3) The user or the client supplies a username and password to the network.
(4) Using 802.1× and EAP, the wireless client and a RADIUS server on the wired LAN perform a mutual authentication through the access point. The RADIUS server sends an authentication challenge to the client. The client uses a one way hash of the user-supplied password to fashion a response to the challenge and sends that response to the RADIUS server. Using information from its user database, the RADIUS server creates its own response and compares that with

the response from the client. Once the RADIUS server authenticates the client, the process repeats in reverse, enabling the client to authenticate the RADIUS server.

(5) When mutual authentication is successfully completed, the RADIUS server and the client determine a WEP key that is distinct to the client and provides the client with the appropriate level of network access, thereby approximating the level of security inherent in a wired switched segment to the individual desktop. The client loads this key and prepares to use it for the logon session.

(6) The RADIUS server sends the WEP key, called a session key, over the wired LAN to the access point.

(7) The access point encrypts its broadcast key with the session key and sends the encrypted key to the client, which uses the session key to decrypt it.

(8) The client and access point activate WAP and use the session and broadcast WEP keys for all communications during the remainder of the session.

Support for EAP and 802.1× delivers on the promise of WEP, providing a centrally managed, standards-based, and open approach that addresses the limitations of standard 802.11 security. In addition, the EAP framework is extensible to wired networks, enabling an enterprise to use a single security architecture for every access method.

Virtual private network security

A VPN-based solution for the wireless LAN security is proposed on the argument that the implementing security at the physical layer is not always practical because a logical connection between two devices may be carried across more than one physical link. Providing end-to-end security between the two end points of a connection is more desirable because it functions independently of underlying data transport. This type of security is best implemented at layer three, which is the network or IP layer. Currently, the most flexible method of providing security at layer three consists of integrating VPN with access points.

Virtual private network provides the means to transmit data securely between two network devices over an unsecure data transport medium. It is commonly used to link remote computers or networks to a corporate server via the Internet, but can also be used to secure wireless networks. Virtual private network works by creating a tunnel on top of a protocol such as IP. Traffic inside the tunnel is encrypted and totally isolated. Virtual private network provides three levels of security: user authentication, encryption, and data authentication:

(1) Authentication ensures that only authorized users over a specific device are able to connect, send, and receive data over the wireless network.

(2) Encryption offers additional protection because it ensures that even if transmissions are intercepted, they cannot be decoded without significant effort.

(3) Data authentication ensures the integrity of data on the wireless network, guaranteeing that all traffic is from authenticated devices only.

Implementation of VPN provides security for the wireless networks that far exceeds the security of unprotected wired devices, and therefore wireless networks can be used for mission-critical challenges without having to worry about security.

Security implementation for VPN: Applying VPN technology to secure a wireless network requires a different approach from what is used on the wired networks. The differences are because:

(1) The repeater function of the wireless access points automatically forwards traffic between wireless LAN stations that communicate together and that appear on the same wireless LAN;
(2) The range of the wireless network usually extends beyond the physical boundaries of an office or enterprise, giving outsiders the means to compromise the network;
(3) The ease with which wireless networking solutions can be produced for different application environments, such as home office, cafes, enterprises, etc. Each of these environments has different security needs and a VPN solution will vary slightly for each of these environments.

In the next section, I present VPN: solutions for an enterprise networking environment. For other environments, please see [2].

Enterprise networks using VPN: In business environments, total security of the wireless network is crucial. This can be impossible to achieve with wireless solutions that rely exclusively on an external server for all VPN functionality. A security hole is created because access must be granted to the wireless network to enable computer users to reach the VPN server and log on. Traffic flow on the wireless network cannot be completely secured. To make effective use of VPN technology, the access points must have their own VPN server, or at the very least be VPN aware. A VPN-aware access point only accepts and forwards VPN traffic to an external VPN server, discarding all other traffic as shown in Fig. 2.3.

Both implementations provide complete security for the network, because the access point will not allow wireless traffic outside of the VPN unless that traffic is to establish a VPN.

2.1.7 Other MAC-based encryption enhancements

In addition to the security solutions described above, MAC-level encryption enhancements are being specified to provide standard, improved encryption

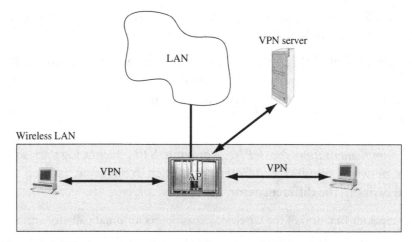

Figure 2.3 Wireless access point with integrated VPN server

and data authentication at the wireless MAC level and to standardize the use
of upper layer authentication. The 802.11i subgroup has proposed these
standards [3] for future use in 802.11 networks. The two possible recommen-
dations of this group are:

(1) A strengthened version of the RC-4/per-frame IV encryption algorithm;
(2) A 128-bit AES encryption algorithm.

Improvements and enhancements that address the shortcomings of WEP have
been identified on the basis of feedback from members of the cryptographic
community. These improvements include:

(1) The addition of a per-packet hash function and IV sequencing rules [4, 5];
(2) The addition of a temporal key derivation algorithm [6];
(3) The addition of rekey mechanisms [6];
(4) The addition of a message authentication code, termed a message integrity code.

Taken together, the enhanced protocol, known as the temporal key integrity
protocol (TKIP), addresses the flaws identified in the current WEP algorithm.
A critical constraint placed on the strengthened WEP algorithm definition is
that it must be able to be implemented and deployed via software upgrades to
the existing base of 802.11 devices.

Encryption at the MAC level can also be strengthened by the proper use
of additional encryption algorithms. The advanced encryption standard
(AES) Rijndael algorithm [7] has been selected by NIST [8] as the next-
generation encryption algorithm, to replace DES and 3DES. Several ways of
using the AES algorithm have been defined. Two of these, AES-OCB [9] and

AES-CBC-MAC [10] are of particular interest for wireless LAN applications. The AES-OCB mode has the attribute of providing both authentication and encryption in one pass through the data. The AES-CBS-MAC algorithm, which combines counter mode encryption with cipher block chaining message integrity and authentication, is termed AES-CCM [11]. The enhanced security standard 802.11i is likely to use one of these enhanced encryption modes.

2.1.8 The IEEE 802.11 security standard – conclusions

Wireless LAN security is a work in progress. The protocols are evolving to meet the needs of serious users. Until the protocols have proven themselves, the best course of action for network engineers is to assume that the link layer offers no security and that one should treat wireless stations as one would treat an unknown user asking for access to network resources over an untrusted network. Policies and resources developed for remote dial-up users may be helpful because of the similarity between a wireless station and a dial-up client. Both are needed for unknown users, who must be authenticated before network access is granted, and the use of an untrusted network means that strong encryption (IPSec, SSL, or SSH) should be required. Although this cautious approach requires much more work than simply throwing up some access points, a conservative approach with several layers of defense is probably the best way to secure an 802.11 network.

After presenting the wireless local area networking security, I shift gear and start discussing the topic of wide area network security. Cellular networks are the prime examples of wide area networks. I have chosen GSM as the model for cellular security, as it has been widely deployed in the world. The second- and third-generation evolutions of GSM are known as GPRS (general packet radio service) and UMTS (universal mobile and terrestrial system), in which GSM security has been enhanced to address well known security vulnerabilities as well as new security schemes proposed to secure data applications.

2.2 Wireless cellular network security

2.2.1 Security for GSM

The GSM wireless network

The global system for mobile (GSM) wireless communications network enables digital wireless duplex communication with data encryption algorithms built in. Before I describe the GSM security, it is important that I describe the GSM network in some detail, albeit briefly.

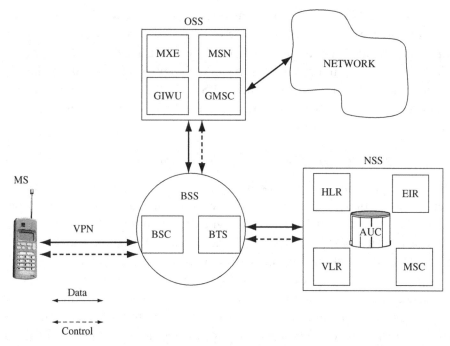

Figure 2.4 The GSM network

The GSM network consists of four major functional components: the mobile station (MS), the network switching system (NSS), the base station system (BSS), and the operation and support system (OSS), as shown in Fig. 2.4.

The mobile station (MS) is the subscriber equipment or the mobile telephone. The *network switching system* (NSS) consists of the home location register (HLR), the equipment identity register (EIR), the visitor location register (VLR), the mobile switching center (MSC), and the authentication center (AUC).

- The HLR stores data about GSM subscribers, including the individual subscriber authentication Key (K_i) for each subscriber identity module (SIM).
- The EIR contains information about the identity of mobile equipment, and prevents calls from stolen, unauthorized, or defective mobile stations.
- The VLR temporarily stores information about roaming GSM subscribers.
- The MSC performs telephony switching functions and is responsible for toll ticketing, network interfacing, and common channel signaling.
- The AUC is a database that contains the international mobile subscriber identity (IMSI), the subscriber authentication key (K_i), and the algorithms that are defined for encryption.

The *base station system* (BSS) connects with the MS over a radio interface link and with the OSS and NSS over cable or fiber links. It consists of the base station controller (BSC) and the base transceiver station (BTS).

- The BSC is the network element that provides all control functions and physical links between the MSC and BTS. It provides functions such as handover, cell configuration data, and control of radio frequency (RF) power levels in base transceiver stations.
- The BTS handles the radio interface to the mobile station. It comprises the radio equipment (transceivers and antenna) that services each cell in the network.

The *operation and support system* (OSS) consists of the message center (MXE), the mobile service node (MSN), the gateway mobile services switching center (GMSC), and the GSM interworking unit (GIWU).

- The MXE provides a short message service (SMS), voice mail, fax mail, email, and paging services.
- The MSN provides mobile intelligent network services.
- The GMSC interconnects two GSM networks.
- The GIWU interfaces to various data networks.

With the completion of this brief description of the GSM system, I am now ready to discuss the security system in place. Here, again, my discussion will be brief. For additional details, see [12].

Security in GSM

Security in GSM wireless networks is established with the use of the following keys:

- The individual subscriber authentication key (K_i) is a 128 bit secret key that is shared between the mobile station and the home location register of the subscriber's home network. It is generated from the SIM utilizing the A8 algorithms.
- The session key (K_c) is a 64 bit ciphering key that is used for encryption of the over-the-air channel. It is generated by the mobile station from the random challenge presented by the GSM network.
- The random challenge (RAND) is a random 32 bit stream generated by the home location register.
- The signed response (SRES) is a 32 bit stream generated by the mobile station and the mobile services switching center.

The GSM wireless network uses the A3 algorithm for authentication, the A5 algorithm for encryption, and the A8 algorithm for key generation. Authentication is an optional procedure at the beginning of a call.

The *A3 algorithm* is implemented in the mobile station's subscriber identity module (SIM). Its task is to generate the 32 bit signed response (SRES). This is accomplished by utilizing a 128 bit random challenge (RAND), which is generated by the home location register (HLR) and the 128 bit individual subscriber authentication key (K_i) from the SIM, or by the HLR.

The *A5 algorithm* is implemented in the mobile station for encryption of the data and is a stream cipher. The stream cipher is initialized with the session key K_c and the frame number. Although the same K_c is used throughout the call, the 22 bit frame number is designed to change during the call, thus creating a unique key stream for every frame. In practice, as long as the mobile service switching center (MSC) does not re-authenticate the mobile station, the same K_c is used for days.

The A5 algorithm has various encryption levels. The A5/0 level utilizes no encryption, A5/1 is used in European countries, A5/2 is a weaker encryption algorithm for use in the United States, and A5/3 is a strong encryption algorithm specially created for the 3rd Generation Partnership Project (3GPP) [12].

The *A8 algorithm* is implemented in the subscriber identity module (SIM) and is responsible for generating the key. From the 128 bit RAND that is received from the MSC, and from the 128 bit key K_i from the SIM, or from the HLR, the A8 generates 128 bits of output, the last 64 of which are the session key K_c. Typically the same K_c is used until the MSC authenticates the mobile station again.

In summary, a GSM duplex conversation is transmitted in a sequence of frames every 4.6 milliseconds. Each duplex frame contains 114 bits in the forward and 114 bits in the reverse direction. A conversation is encrypted by the session key K_c. For each frame, K_c is mixed with a publicly known frame counter, F_n, and the result serves as the initial state of a pseudo-random generator that produces 228 bits. These bits are XORed by the two parties with the $114 + 114$ bits of plain text to produce $114 + 114$ bits of cipher text.

2.2.2 Security in GPRS networks

Before discussing security for a GPRS network, it is logical to discuss the GPRS network in brief; GPRS networks are described in detail in [13]. General Packet Radio Service (GPRS) is a second-generation cellular system, which has been created by enhancing the GSM system to support data services. In GPRS, the voice call is handled by the GSM network and the data calls are handled by a separate overlay network that has a few new nodes known as GPRS support nodes (GSNs). A GPRS logical architecture is shown in Fig. 2.5.

A GPRS support node contains the functionality required to support GPRS. Two GSNs have been created for GPRS. These are:

- The gateway GPRS support node (GGSN) is the node that is accessed by the packet data network for packet forwarding and routing. It contains routing information

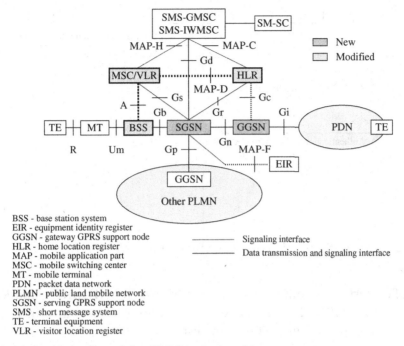

BSS - base station system
EIR - equipment identity register
GGSN - gateway GPRS support node
HLR - home location register
MAP - mobile application part
MSC - mobile switching center
MT - mobile terminal
PDN - packet data network
PLMN - public land mobile network
SGSN - serving GPRS support node
SMS - short message system
TE - terminal equipment
VLR - visitor location register

Figure 2.5 Overview of the GPRS logical architecture

for attached GPRS users. The routing information is used to tunnel packets to the mobile station's current point of attachment to the network, i.e., the serving GPRS support node (SGSN). The interfaces that GGSN supports are shown in Fig. 2.5. The GGSN is the first point of interconnection with a packet data network.

- The serving GPRS support node (SGSN) is the node that is serving the mobile station. When a node requires service from the GPRS, it has first to attach to the network. "Attach" is a GPRS procedure, and means establishing a logical connection with the network using wireless channels. The SGSN is responsible for radio resourcing and security, as well as mobility management of mobile terminals.

Three types of security function are supported in GPRS networks: authentication and service request validation, user identity confidentiality, and user data confidentiality. In the following, I briefly cover these three security mechanisms.

Authentication of subscriber

The GPRS uses the authentication procedure already defined for GSM with the distinction that the procedures are now executed in the SGSN and not in MSC. The GPRS authentication procedure performs subscriber authentication, or selection of the ciphering algorithm and the synchronization of the

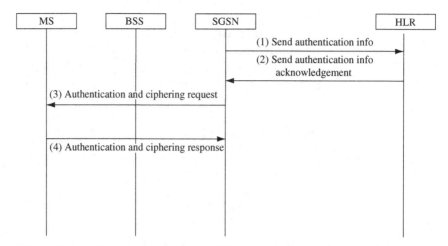

Figure 2.6 Authentication procedure for GPRS

start of ciphering, or both. Authentication triplets (RAND, SRES, and K_c) are stored in the SGSN. The authentication procedure is illustrated in Fig. 2.6. Each step is explained in the following list.

(1) If the SGSN does not have authentication triplets, it sends a message to HLR for getting the triplets;
(2) The HLR responds to the SGSN in a separate message with a triplet;
(3) The SGSN sends an authentication and ciphering request message to the MS;
(4) The MS responds with an authentication and ciphering response (SRES) message.

The mobile station starts ciphering after sending the authentication and ciphering response message. The SGSN starts ciphering when a valid authentication and response is received from the MS. The ciphered information from the SGSN and, MS is compared and, if found identical, the MS is authenticated.

For user identity confidentiality, GPRS uses the authentication procedure described above. The GPRS uses the concept of temporary mobile station identity (TMSI) to authenticate a mobile station, which is allocated by SGSN as part of the attach procedure [13]. The procedures for user data confidentiality and MS identity check are identical to GSM with the exception that the procedures are executed from SGSN, and logical link control (LLC) frame numbers are used instead of TDMA frame numbers, which are not known.

2.2.3 Security for UMTS

Before discussing the UMTS security, it is logical that I briefly go over UMTS network architecture. Figure 2.7 provides a logical UMTS network architecture.

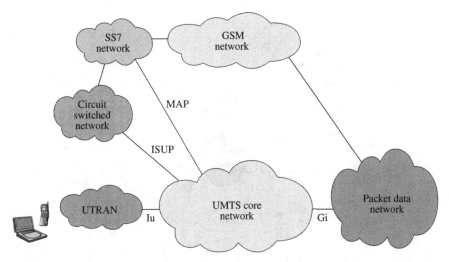

Figure 2.7 The UMTS network architecture

Figure 2.7 shows a high-level view of the UMTS network architecture, which consists of six main subsystems and several major interfaces. These six components are the UTRAN (UMTS terrestrial radio access network), core network, SS7 network, GSM/GPRS network, circuit-switched network, and the external packet data network. Major interfaces between the entities are also given. The core network comprises a circuit-switched (CS) domain for providing voice and the circuit-switched data services and a packet-switched (PS) domain for providing packet-based services. Figure 2.8 depicts a logical architecture of the UMTS by showing the CS domain on the left and the PS domain on the right. The radio access network for both the domains is UTRAN, which is composed of a set of radio network subsystems (RNS). The radio network systems are, in turn, composed of two main logical elements: node B and a radio network controller (RNC). The RNS is responsible for the resources and the transmission and reception in a set of cells where a set of cells (sector) is one coverage area served by a broadcast channel.

The radio network controller is responsible for the use and allocation of all the radio resources of the radio network system to which it belongs. A radio network controller can logically be split into two entities: the first entity controls the base stations (base station controller), and the second entity is responsible for traffic processing.

The base station controller is mainly in charge of the allocation and usage of all radio resources with the aim of hiding the details from the core network. The traffic-processing unit mainly handles the user voice and packet

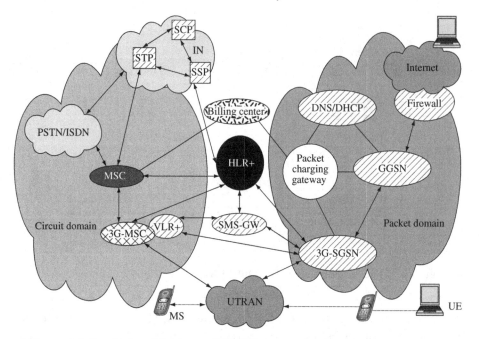

Figure 2.8 Logical architecture of UMTS core network

data traffic, performing the actions on the user data streams that are necessary to access the radio bearers. Node B is mainly responsible for radio transmission and reception in one or more cells to and from the user equipment (UE).

The UMTS Iu interface is an open logical interface that interconnects one UTRAN to the core network. The Iu interface is described in detail in [14]. In the following, I briefly describe the functions of some of the nodes in the circuit and packet domains. For details, please see [15].

Circuit-switched domain

The 3G-MSC is the main core network element for providing circuit-switched services. The 3G-MSC provides the necessary control and the corresponding signaling interfaces, e.g., SS7, MAP, and ISDN. The 3G-MSC may also provide gateway functionality for interconnecting to external networks like PSTN and ISDN. The key functionality provided by the MSC includes: mobility management; call management; supplementary services, such as call forwarding; circuit-switched data services; and support of SS7, MAP, and RANAP interfaces to complete originating or terminating calls in the network in interaction with other entities of a mobile network, e.g., HLR, short message services, and VLR (visitor location register).

Packet-switched domain

The important network nodes that form the packet domain are:

3G-SGSN: The 3G-SGSN is the main core network element for packet services. It provides the necessary control functionality towards both the UE and the GGSN. It also provides the appropriate signaling and data interfaces. These include connection of an IP-based network towards the GGSN, SS7 towards the HLR, and TCP/IP or SS7 towards the UTRAN.

Functionalities provided by the 3G-SGSN include: session management; support for Iu, Gn, and MAP interfaces; short message services, mobility management, subscriber database functionality, and charging.

GGSN: provides inter-working with the external packet-switched network and is connected with the SGSN via an IP based network. The GGSN may optionally support an SS7 interface towards the HLR, which is used to handle mobile terminated packet sessions.

Functionalities provided by the 3G-GGSN include: maintenance of information on mobile location at SGSN level; the formation of a gateway between the UMTS packet network and external data networks (e.g., IP, X.25), the provision of gateway-specific access methods to intranets (e.g., PPP termination), user data screening and security, and support for charging for the external data network usage.

Firewall: This entity is used to protect the operator's backbone data network from attack from external packet data networks.

The DNS and DHCP servers: The DNS server is used, as in any IP network, to translate host names into IP addresses. A DHCP (dynamic host configuration protocol) server is used to manage the allocation of IP configuration information by automatically assigning IP addresses to systems configured to use DHCP. Additionally, the GGSN may have to allocate a dynamic address to the UE upon PDP context activation.

Packet charging Gateway: This provides a mechanism to transfer charging information from the SGSN and GGSN nodes to the network operator's chosen billing center.

Billing center: This collects CDRs (call detail records) and produces customer-billing information.

SMS-GMSC: This supports short message service via a gateway MSC (GMSC) for packet and circuit mode messages.

In basic GSM, security is concentrated on the radio path security, which is implemented in the access network part of the network. In UMTS networks, security is a broader topic. Radio path security is important, but in addition to this, security must be provided in other vulnerable areas as well, e.g., security of connections inside the UMTS network, and between the networks controlled by different mobile network operators.

The radio access technology changes from TDMA in GSM and GPRS to WCDMA for UMTS. However, access security requirements do not change from GSM. In UMTS, as in GSM, it is also required that end users of the system are authenticated. The GSM-based authentication procedures have been carried over to UMTS with slight enhancements to overcome some of the GSM security vulnerabilities. Some well known vulnerabilities of GSM are:

(1) Sensitive control data, e.g., keys used for radio interface ciphering, are sent between different network elements without ciphering;
(2) Some parts of the security architecture are kept secret, e.g., the cryptographic algorithms: this precludes scrutiny of such algorithms by the public with regard to their vulnerabilities, and subsequent attempts to strengthen them;
(3) Vulnerability of radio interface ciphering keys due to brute force attacks for breaking them.

The UMTS implements solutions in the security architecture to overcome these limitations in the access part of the network. Some of the new solutions are:

● Mutual authentication of the users and the networks;
● Use of temporary identities;
● Radio access network encryption;
● Protection of signaling messages inside UTRAN.

In this section, I briefly cover these solutions. For additional details, please refer to [12].

Mutual authentication

There are three entities involved in the authentication mechanism of the UMTS system:

(1) Home network;
(2) Serving network (SN);
(3) Mobile Station with USIM card (USIM is the UMTS version of GSM SIM card).

The basic idea is that the serving network (SGSN) checks the subscriber's identity (as in GSM) by a so-called challenge-and-response technique while the mobile station checks that the serving network has been authorized by the home network to do so. The latter part is a new feature in UMTS, compared with GSM, and through it the mobile station can check that it is connected to a legitimate network.

The cornerstone of the authentication mechanism is a master key (K_i) that is shared between the USIM of the user and the home network database. This is a

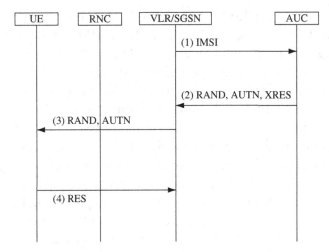

Figure 2.9 User authentication in UMTS

permanent secret with a length of 128 bits. The key (K_i) is never transferred out from the two locations. The keys for mutual authentication and the keys for encryption and integrity checking are derived using the master key during every authentication event and the derived keys also have the same length of 128 bits.

Now I am ready to discuss the UMTS authentication at a higher level. The authentication procedure can be started after the user is identified in the serving network. The identification occurs when the identity of user, i.e., its permanent identity (IMSI) or temporary identity (TMSI), has been transmitted to the VLR or SGSN. Then the VLR or SGSN sends an *authentication data request* to the authentication center (AuC) in the home network. The authentication center contains master keys of users and, based on the knowledge of the IMSI, the AuC is able to generate authentication vectors for the user. The generation process contains execution of several cryptographic algorithms, which are described in [12] in more detail. The generated vectors are sent back to the VLR or SGSN in the authentication data response. This process is depicted in Fig. 2.9. These messages are signaling messages.

In the serving network, one authentication vector is needed for each authentication instance, i.e., for each run of the authentication procedure. This means that the signaling between the SN and the AuC is not needed for every authentication event and it can, in principle, be done independently of the user actions after the initial registration. The VLR or SGSN may fetch new authentication vectors from the AuC well before the number of stored vectors runs out.

The VLR or the SGSN of the serving network sends a user authentication request to the terminal. This message contains two parameters from the authentication vector, called RAND and AUTN. These parameters are transferred into the USIM of the mobile station. The USIM contains the master key K_i and, using it with the parameters RAND and AUTN as inputs, USIM generates an authentication vector similar to that in AuC. The USIM verifies that the AUTN that it received from the AuC is the same as it just generated. If the result is positive, then the computed parameter RES is sent back to the VLR or SGSN in the *user authentication response*. Now the VLR or SGSN is able to compare the user response, RES, with the expected response, XRES, which is part of the authentication vector. In the case of a match, the MS is authenticated. The keys for radio access network encryption and integrity protection, namely CK and IK, are created as a by-product in the authentication process. These temporary keys are included in the authentication vector and, thus, are transferred to the VLR or SGSN. These keys are later transferred to the RNC in the radio access network when the encryption and integrity protection are started. The USIM, however, is also able to compute CK and IK after it has obtained the RAND and verified it through the AUTN. Temporary keys are later transferred from USIM to MS for use in the encryption and integrity protection algorithms.

Encryption in the UTRAN

Once the user and the network have authenticated each other they may begin secure communication. As discussed, a cipher key, CK, is shared between the core network and the terminal after a successful authentication event. Before encryption can begin, the communicating parties have to agree on the encryption algorithm. The UMTS defines such an algorithm for its release 99 [14]. The encryption and decryption takes place in the terminal and the radio network controller (RNC), which is part of the UTRAN. This means that the cipher key, CK, has to be transferred from the core network to the radio access network. The core network uses a special layer-three message to transmit the key to RNC. After the RNC has obtained the key it can switch on the encryption by informing the terminal in another special layer three message. UTRAN encryption is described in detail in [12] and [15].

2.3 Bluetooth or IEEE 802.15 security

Bluetooth is a short-range wireless communication standard that enables personal area networking among a wide variety of personal devices ranging from laptops to cell phones, computers to printers, personal digital assistants to wireless headsets, and many other devices and applications. An excellent introduction to Bluetooth is given in [16] and [17], and the interested reader should reference these. In this section I briefly describe the security aspects of the Bluetooth devices. Please see [16] for the details of Bluetooth security.

Like any other wireless network that is prone to signal interception and subsequent decoding, Bluetooth is no exception, but to provide for a secure communication among the Bluetooth devices, the standard provides protection from eavesdropping or falsifying the origin of messages, which is known as spoofing. Bluetooth applications may choose from several levels of error correction encoding techniques to facilitate reliable communication. The technology also provides for several levels of secure communication by stipulating protocols and procedures for authentication, authorization, and encryption at the hardware level as well as the software level. Because of its strong security features and interface management procedures, Bluetooth enables concurrent networks in the same geographic space, allowing devices to participate in different networks at the same time.

The main security features that a Bluetooth device can have are:

(1) A challenge-response routine for authentication, which prevents spoofing and unwanted access to critical data and functions;
(2) Stream cipher for encryption, which prevents eavesdropping and maintains link privacy;
(3) Session key generation – session keys can be changed at any time during a connection.

Bluetooth devices can use the following entities in the security algorithms they execute to provide secure communication:

(1) The 48 bit Bluetooth device address is a public entity unique for each device and can be obtained through the inquiry procedure;
(2) The 128 bit private user key is a secret entity that is derived during initialization and is never disclosed;
(3) A 128 bit random number is derived from a pseudo-random process in the Bluetooth unit, generating a different number for each new transaction.

These are link-layer functions for providing security for Bluetooth devices, but frequency hopping and the limited transmission range also helps to prevent eavesdropping.

2.4 Summary and further reading

This chapter provided a basic introduction to wireless security for WAN, LAN, and PAN environments in brief. A detailed treatment of all the schemes in one chapter is not feasible. For additional information, interested readers should consult the corresponding references provided in this chapter.

2.5 References

[1] IEEE Standard 802.11, 1999 Edition (R2003) (ISO/IEC 8802–11:1999), *IEEE Standard for Information Technology – Telecommunications and Information Exchange between Systems – Local And Metropolitan Area Network Specification Requirements – Part II. Wireless LAN Medium Aceess Control (MAC) and Physical Layer (PHY) Specifications*, 1999.

[2] B. Bing, *Wireless Local Area Networks*, John Wiley and Sons, 2002.

[3] IEEE 802.11 Study Group, http://groupr.ieee.org/groups/802/11.

[4] R. Housely and D. Whiting, *Temporal Key Hash*, IEEE Standard 802.11-01/550.

[5] R. Housely, D. Whiting, and N. Ferguson, *Alternative Temporal Key Hash*, IEEE Standard 802.11i, 11-02-282r0.

[6] T. Moore and C. Chaplin, *TGi Security Overview*, IEEE Standard 802.11i, 10-02-114r1.

[7] www.esat.kuleuven.ac.be/~rijmen/rijndael.

[8] National Institute of Standards and Technology, www.nist.gov.

[9] P. Rogaway, www.cs.ucdavis.edu/~rogaway.

[10] S. Frankel, R. Glenn, and S. Kelly, "*The AES-CBC Cipher Algorithm and Its Use with IPSec*, www1.ietf.org/mail-archive/web-dd/ietf-announce-old/Current/msg26234.html, 2003.

[11] R. Housely, D. Whiting, and N. Fergusen, *AES-CTR-Mode-with-CBC-MAC*, 80211-02-001r1.

[12] 3rd Generation Partnership Project: Technical Specification Group (Services and System Aspects) *3G Security; Specifications of A5/3 Encryption Algorithms for GSM and ECSD, and the GEA3 Encryption Algorithm for GPRS; Document 1:A5/3 and GEA3 Specifications (Release 6)*, 3GPP TS 55.216, ver. 6.2.0, 2003.

[13] ETSI, *Digital Cellular Telecommunications System (Phase 2+), General Packet Radio Service (GPRS); Service Description; Stage 2, GSM 03.60, ver. 6.4.0*, 1997.

[14] *3GPP Technical Specifications*: www.3gpp.org.

[15] H. Kaaranen, A. Ahitiainen, L. Laitinen, S. Naghian, and V. Niemi, *UMTS Networks: Architecture, Mobility, and Services*, John Wiley and Sons, 2001.

[16] C. Bisdikian, P. Bhagwat, B. Gaucher, *et al.*, "WiSAP – A wireless personal access network for handheld computing devices," *IEEE Personal Commun.*, vol. 5, no. 6, 1998, pp. 18–25.

[17] Bluetooth SIG, *Bluetooth – Specification of the Bluetooth System*, vol. 1, core, ver. 1.1.

3

Threats and attacks

Ad hoc networks are vulnerable not only to attacks from outside but also from within. Moreover, these attacks can be active as well as passive. In this chapter, I will discuss the various possible attacks on wireless ad hoc networks; this will then facilitate the discussion about designing the security schemes for such attacks in subsequent chapters. The significance of the various security needs discussed in Chapter 1 now comes to the fore, since any attack essentially disrupts either the operational mechanisms, or the security mechanisms including the security apparatus.

3.1 Attack classification

Ad hoc networks are typically subjected to two different levels of attacks. In the first level of attack, the adversary focusses on disrupting the basic mechanisms of the ad hoc network, such as routing, which are essential for proper network operation, and in the second level of attacks, the adversary tries to damage the security mechanisms employed by the network, such as key management schemes or cryptographic algorithms in use. This can be one way of classifying attacks. Alternatively, attacks against ad hoc networks can be classified into two groups in a different way:

(1) *Passive attacks* which involve only *eavesdropping* on the data that is being communicated in the network. Examples of passive attacks include covert channels, traffic analysis, sniffing to compromised keys, etc., and
(2) *Active attacks* which involve specific actions performed by adversaries, for instance, the replication, modification, and deletion of exchanged data among the nodes.

Adversaries attempt to change the behavior of the operational mechanisms in active attacks while they are subtle in their activities in passive attacks.

The information gathered as a result of data sniffing during a passive attack may subsequently form a basis for an active attack. Attacks are also classified according to the facilities used by the attackers. For example an attack launched by a remote adversary will be classified as *external* whereas an attack launched by one of the nodes which is part of the network will constitute an internal attack.

External attacks are typically active attacks that try to cause congestion in the network, propagate incorrect routing information, prevent services from working properly, or shut down the network completely. *External attacks* can typically be prevented by using standard security mechanisms, such as firewalls, encryption, and other cryptography based algorithms, etc. *Internal attacks* are typically more severe attacks, since malicious insider nodes already belong to the network as an authorized party and are thus protected by the security mechanisms of the network and its services. Thus, such malicious insiders, who may even operate in a group, may use the standard security means to actually protect their own attacks.

In this chapter, I will discuss some important attacks that could be launched against the ad hoc network as well as individual nodes.

3.2 Denial of service (DoS)

The denial of service threat produced either by an unintentional failure in the system or a malicious action forms a severe security risk in any distributed system. The classical way to create a DoS attack is to flood any centralized resource so that it no longer operates correctly or crashes. But in ad hoc networks, this may not be an applicable approach, due to the distribution of responsibility as well as the lack of a centralized resource. Radio jamming and battery exhaustion are two other ways in which service can be denied to other nodes and users. A distributed DoS attack is an even more severe threat. If the attackers have enough computing power and bandwidth to operate with, smaller ad hoc networks can be crashed or congested rather easily. Compromised nodes may be able to reconfigure the routing protocol or a part of it, such that they can send routing information very frequently, thus causing congestion and preventing nodes in gaining the latest information about the changed topology of the network. If the presence of compromised nodes and the compromised routing are not detected, the consequences to the network are severe, as the network may seem to operate normally to the other nodes. This kind of invalid operation of the network initiated by malicious nodes is called a *Byzantine failure*. For example, a compromised node could participate in a session but simply drop a certain number of packets, which may lead to degradation in the quality of service being offered by the network. In summary, some of the examples of Denial of Service attacks are:

- *SYN flooding* In this type of DoS attack, the adversary sends a large number of SYN packets to a victim node, spoofing the return address of the SYN packets. On receiving the SYN packets, the victim node sends back acknowledgement (SYN-ACK) packets to nodes whose addresses have been specified in received SYN packets and awaits for ACKs from the senders, which never arrive. If sufficient connections are established among multiple senders and the victim, it is likely that its memory resources may be exhausted (table overflow), owing to the currently open connections and the victim cannot now accept a new legitimate request for a connection.
- *Jamming* This type of DoS attack is initiated by a malicious node after determining the frequency of communication used by the receiver and using the same frequency to send data to the receiver thereby interfering with its operation. Frequency hopping is an established technique to get around jamming attacks.
- *Distributed denial of service attack* This type of attack is launched by a group of compromised nodes who are part of the same network and who collude together to bring the network down or seriously affect its operation.

3.3 Impersonation

Impersonation attacks form a serious security risk at all levels of ad hoc networking. If proper authentication of parties is not supported, compromised nodes may be able to join the network, send false routing information, and masquerade as some other trusted nodes. A compromised node may get access to the network management system of the network and may start changing the configuration of the system as a super-user who has special privileges. At the service level it is then possible that a malicious party could have its public key certified even without proper credentials. A malicious party may be able to masquerade itself as any of the friendly nodes and give false orders or status information to other nodes. Impersonation threats are mitigated by applying strong authentication mechanisms in contexts where a party has to be able to trust the origin of data it has received or stored. Most often, this means application of digital signatures or keyed fingerprints over routing messages, configuration or status information, or exchanged payload data of the services in use in as many layers of the protocol stack as possible. Digital signatures implemented with public-key cryptography are a problematic issue within ad hoc networks, as they require an efficient and secure key management service and require relatively more computation power. Thus, in many cases, lighter solutions, such as the use of keyed hash functions, or a priori negotiated and certified keys and session identifiers, are needed. They do not, however, remove the demand for secure key management or proper confidentiality protection mechanisms.

Two well known impersonation attacks, Sybil and Trust, are discussed below.

3.3.1 Sybil attack

In the Sybil attack [1, 2] a malicious node behaves as if it were a larger number of nodes (instead of one) by impersonating other nodes or simply by claiming false identities. In the worst case, a Sybil attacker may generate an arbitrary number of additional node identities, using only one physical device. The additional identities that the node acquires are called Sybil nodes. There are three possible dimensions in which a Sybil attack can be launched. Each of these dimensions is discussed below.

(1) Direct or indirect communication Direct communication In this case, one way to perform the Sybil attack is for the Sybil nodes to communicate directly with legitimate nodes. When a legitimate node sends a radio message to a Sybil node, one of the malicious devices listens to the message. Likewise, messages sent from Sybil nodes are actually sent from one of the malicious devices.

Indirect communication In this type of attack the communication between a Sybil node and a legitimate node is indirect, i.e., via another malicious node. In other words, legitimate nodes are not able to communicate directly with the Sybil nodes. Messages sent to a Sybil node are routed through one of these malicious nodes, which then pass it on to a Sybil node.

(2) Fabricated or stolen identities A Sybil node has two options for getting an identity to itself. The first option is that it fabricates a new identity for itself. The second option is that it steals an identity of a legitimate node. While using the first option, a Sybil node can create an arbitrary and random 32 bit integer number as an identifier, if the network, for example, is using 32 bit identifiers for nodes.

A Sybil node has, somehow, to find legitimate identities for communicating with other legitimate nodes, and one option is if it can somehow steal these. The easiest way to get an identity is to get the identity of an impersonated node, if such a node exists in the network. The identity theft can remain undetected if the impersonated node is destroyed or temporarily disabled from the network. If the range of legitimate identities is limited by some security mechanisms then identity fabrication can be difficult.

(3) Simultaneity Simultaneous The attacker may try to have all his or her Sybil identities participate in the network at once. While a particular hardware entity can only act as one identity at a time, it can cycle through these identities to make it appear that they are all present simultaneously.

Non-simultaneous In this type of attack, some node identities are used in one time interval, and others are used in the next time interval. Also, if the attackers have several compromised nodes, then these nodes can swap their identities on a periodic basis and remain undetected.

Known examples of Sybil attacks

Here are some of the known applications of Sybil attacks for wireless ad hoc and sensor networks:

(1) *Routing* Sybil attacks have been shown to be effective against routing protocols in ad hoc and sensor networks. The multi-path and disparity routing algorithms are particularly vulnerable if the path consisting of multiple segments goes through a single malicious node presenting several Sybil node identities. This attack can also affect geographical routing protocols when a Sybil node appears in several locations at once instead of appearing in one place.

(2) *Data aggregation* Sensor networks use query protocols, which compute aggregates of sensor readings within the network to conserve energy, rather than returning individual sensor readings. If there are a small number of malicious nodes reporting erroneous sensor readings, then they may not be able to affect the aggregate reading by a wide margin. But by using a Sybil attack, a node may be able to contribute to the aggregate many times, thereby affecting the aggregate sensor reading.

(3) *Voting* In certain applications, sensors can be used to perform voting, in order to facilitate decision making. Because of the ability of Sybil nodes to replicate identities, such nodes can affect the outcome of any vote.

(4) *Misbehavior detection* The presence of Sybil nodes in a network may make it difficult to identify a misbehaving node. An attacker with many Sybil nodes could "spread the blame" by not having any one Sybil identity misbehave enough for the system to take action. Also, if the action taken is to revoke the offending node, the attacker can simply continue using new Sybil identities to misbehave, never getting revoked himself or herself.

(5) *Fair resource allocation* In networking, resources are often shared among the nodes and often allocated on a per node basis. For example, a wireless channel using TDMA MAC may assign the same channel to different users for short intervals of time (time slots). The Sybil attacker can disrupt the fair allocation of resources by assigning a resource to the same node several times by changing its identity.

This section is primarily focussed on introducing Sybil attack and its characteristics without delving into mechanisms for its detection. For defense mechanisms that can be used against the Sybil attacks, see [3].

3.3.2 *Trust attack*

A trust attack is another type of impersonation attack. In simple security applications, in which the goal is to protect a given message or an item from passive or active attacks, user trust can be established as an authentication procedure between a system and a user. But there are applications which require multiple security levels. For example military applications have information that is categorized as unclassified (U), confidential (C), secret (S), or top secret (TS), and each type of information can require a set of authentication rules that have some sort of hierarchical structure, called a trust hierarchy. A trust hierarchy is basically an explicit representation of trust levels that reflects organizational privileges. It associates a number with each privilege level, to reflect the security, importance, or capabilities of the mobile node and also of the paths. Attacks on the trust hierarchy can be broadly classified as outsider attacks and insider attacks, based on the trust value associated with the identity or the source of the attack. What is also needed is a binding between the identities of the users with the associated trust levels. Without this binding, any user can impersonate anybody else and obtain the privileges associated with higher trust levels. To prevent this, stronger access control mechanisms are required (authentication, authorization, and accounting or AAA). To force the nodes and users to respect the trust hierarchy, cryptographic techniques, e.g., encryption, public key certificates, shared secrets, etc., can be employed. Traditionally, strong authentication schemes are used to combat outsider attacks. The identity of a user is certified by a centralized authority, and can be verified using a simple challenge–response protocol. Insider attacks are launched by compromised users within a protection domain or trust level. Routing protocol packets in existing ad hoc networks do not carry authenticated identities or authorization credentials, and hence compromised nodes can potentially cause a lot of damage. Insider attacks, in general, are hard to prevent at the protocol level. Some techniques to prevent insider attacks include secure transient associations, and tamper-proof and tamper-resistant nodes. Tamper-proof and tamper-resistance concepts are discussed in Section 3.7.

3.4 Disclosure

Any communication must be protected from eavesdropping whenever confidential information is exchanged. Also, the critical data that the nodes store must be protected from unauthorized access. In ad hoc networks, such information can include almost anything, e.g., the specific status details of a node, the location of nodes, private or secret keys, passwords, and so on. Sometimes the control data are more critical for security than the traffic data. For instance, the routing directives in packet headers such as the identity or location of the nodes can be more valuable than the application-level messages. The identities of the observed nodes, traffic patterns around a node, or the detected radio transmissions that a node generates, may be just the information an adversary needs to launch a well targeted attack.

3.5 Attacks on information in transit

In addition to exploiting vulnerabilities related to the protection and enforcement of the trust levels, compromised or enemy nodes can utilize the information carried in the routing protocol packets to launch attacks. These attacks can lead to corruption of information, disclosure of sensitive information, theft of legitimate service from other protocol entities, or denial of network service to protocol entities. Threats to information in transit include:

Interruption The flow of routing protocol packets, especially route discovery messages and updates, can be interrupted or blocked by malicious nodes. Attackers can selectively filter control messages and updates, and force the routing protocol to behave incorrectly.

Interception and subversion Routing protocol traffic and control messages, e.g., "Keep alive" and "Are you up?" messages can be deflected or rerouted.

Modification The integrity of the information in routing protocol packets can be compromised by modifying the packets themselves. False routes can be propagated, and legitimate nodes can be bypassed.

Fabrication False route and metric information can be inserted into legitimate protocol packets by malicious insider nodes.

3.6 Attacks against routing or network layer

Attacks against routing are basically of two types: *internal* and *external*. External attacks can again be classified as active or passive. In this section, the various kinds of routing attacks are briefly discussed.

3.6.1 Internal attacks

An internal attack is a more severe kind of threat to ad hoc networks. The attacker may broadcast wrong routing information to other nodes within the network. A compromised node can be categorized as a source of internal attacks. Detecting modified information in routing protocols is inherently difficult because compromised nodes are able to generate valid signatures using their private keys. Also differentiating between modifications in data due to an actual attack or due to wireless link impairment may be difficult.

3.6.2 *External attacks*

External attacks on routing can be divided into two categories: *passive* and *active*. *Passive attacks* involve unauthorized "listening" to the routing packets. This might be an attempt to gain routing information from which the attacker could extrapolate data about the positions of each node in relation to the others. For example, an attacker that eavesdrops on all the routing updates transmitted in a certain part of the ad hoc network can begin to piece together proximity information of the nodes. For example, it can figure out which nodes are close together (one or two hops apart) and which nodes are far from each other (many hops apart). In a *passive* attack, the attacker does not disrupt the operation of a routing protocol but only attempts to discover valuable information by listening to the routed traffic. This type of attack is usually impossible to detect, which makes defending against such attacks difficult. Furthermore, routing information can reveal relationships between nodes or disclose their IP addresses. If a route to a particular node is requested more often than to other nodes, the attacker might expect that the node is important for the functioning of the network, and may possibly conclude that disabling it could bring the entire network down. Other interesting information that is disclosed by routing data is the location of nodes, as discussed above. Even when it might not be possible to pinpoint the exact location of a node, one may be able to discover information about the network topology. It is worth noting that in an IP network, one cannot defend against these attacks, for example, by only using IPsec. The packets still have most of their IP headers in plaintext, and it may not even be feasible to have symmetric keys distributed to every node in a network.

Active attacks on the network from outside sources are meant to degrade or prevent message flow between the nodes. Active external attacks on the ad hoc routing protocol can collectively be described as denial-of-service attacks, causing a degradation or complete halt in communication between nodes. One type of attack involves insertion of extraneous packets into the network in order to cause congestion. A more subtle method of attack involves intercepting a routing packet, modifying its contents, and sending it back into the network. Alternatively, the attacker can choose not to modify the packet's contents but rather to replay the packet back to the network at different times, introducing outdated routing information to the nodes. The goal of this form of attack is to confuse the routing nodes with conflicting information, delaying packets, or preventing them from reaching their destination. To perform an active attack, the attacker must be able to inject arbitrary packets into the network. The goal may be to attract packets destined to other nodes to the attacker for analysis or just to disable the network. An active attack can sometimes be detected and this makes active attacks a less inviting option for most attackers. Some types of active attacks that can usually be easily performed against an ad hoc network are described below:

(1) *Black hole* In this attack, a malicious node uses the routing protocol to advertize itself as having the shortest path to the node whose packets it wants to intercept. In a flooding-based protocol, the attacker listens to requests for routes. When the attacker receives a request for a route to the target node, the attacker creates a reply consisting of an extremely short route. If the malicious reply reaches the requesting node before the reply from the actual node, a forged route gets created. Once the malicious device has been able to insert itself between the communicating nodes, it is able to do anything with the packets passing between them. It can choose to drop the packets to perform a denial-of-service attack, or alternatively use its place on the route as the first step in a man-in-the-middle attack.

(2) *Routing table overflow* In this attack, the attacker attempts to create routes to non-existent nodes. The goal is to create enough routes to prevent new routes from being created or to overwhelm the protocol implementation. Proactive routing algorithms attempt to discover routing information even before it is needed, while a reactive algorithm creates a route only once it is needed. An attacker can simply send excessive route advertizements to the routers in a network. Reactive protocols, on the other hand, do not collect routing data in advance.

(3) *Sleep deprivation* Usually, this attack is practical only in ad hoc networks where battery life is a critical parameter. Battery-powered devices try to conserve energy by transmitting only when absolutely necessary. An attacker can attempt to consume batteries by requesting routes, or by forwarding unnecessary packets to the node using, for example, a black hole attack. This attack is especially suitable against devices that do not offer any services to the network or offer services only to those who have some special credentials. Regardless of the properties of the services, a node must participate in the routing process unless it is willing to risk becoming unreachable to the network.

(4) *Location disclosure* A location disclosure attack can reveal something about the locations of nodes or the structure of the network. The information gained might reveal which other nodes are adjacent to the target, or the physical location of a node. Routing messages are sent with inadequate hop-limit values and the addresses of the devices sending the ICMP error messages are recorded. In the end, the attacker knows which nodes are situated on the route to the target node. If

the locations of some of the intermediary nodes are known, one can gain information about the location of the target as well.

(5) *Wormhole attack* In this attack, an attacker receives packets at one location in the network and tunnels them (possibly selectively) to another location in the network, and from there the packets are resent into the network. This tunnel between two colluding attackers is referred to as a wormhole. It could be established through a single long range wireless link or even through a wired link between the two colluding attackers. Owing to the broadcast nature of the radio channel, the attacker can create a wormhole even for packets not addressed to itself. If wormholes are created purely for packet relaying purposes, then wormholes are harmless, provided the attacker has no malicious intentions. If the attacker is malicious, then the wormhole can compromise the security of the network. If there are no security mechanisms deployed in the network to prevent wormhole attacks, the existing ad hoc routing protocols are not likely to discover valid routes for packet forwarding.

3.7 Node hijacking

The power at which a signal is transmitted at any given frequency is not policed by any single entity. Communication channels coexist in the same geographical area by cooperating and following some standards. However, there is no regulatory authority to enforce the power-level policy. This means that a malicious node can capture the channel at any time for long periods of time without letting a well behaved authorized user communicate on the channel. A malicious node may also pose as a base station and encourage mobiles to connect to it and collect data (passwords, secret keys, logon names, etc.) and information from these nodes. This is an example of a node hijacking where a legitimate base station has been hijacked by a malicious node.

A hijacking attack is perpetrated remotely by abusing routing protocols and, as a result, leads to detouring of messages or *"route hijacking."* Here the hijacker modifies the routing information in an effort to hijack traffic to and from selected nodes. Using trustworthy nodes or *tamper-proof* nodes to route traffic may be a solution to *node hijacking* or *route hijacking attacks*. Another well known approach to overcoming route hijacking is by employing reputation-based control. In this approach, routers may keep "reputation tables" or "reputation caches" that list nodes they trust. The routing protocols can be made immune to node or route hijacking attacks by enabling routers to keep and possibly exchange reputation-based information. Routers can then use this information to resolve conflicting updating information, and to determine what control messages to handle and act on.

3.7.1 Tamper resistance and tamper proofing

We know that the typical security solutions that are proposed for wireless ad hoc networks consist of cryptography-based algorithms, such as symmetric cipher, public-key cipher, and hash functions. However, these cryptographic algorithms alone cannot ensure security, since most ad hoc systems present attackers with an abundance of opportunities to observe or interfere with their implementations, and, in the process, to compromise their theoretical strength. To provide security in such circumstances, it is imperative that ad hoc nodes be *tamper resistant*. A tamper-resistant design refers to the process of designing an ad hoc node (AN), architecture, and implementation that is resistant to such attacks. From the hardware perspective, an ad hoc node resembles an embedded system.

Two areas of security that are pertinent to AN security in the context of tampering are related to data integrity and confidentiality. Data integrity ensures that the data in an ad hoc node has not been deleted or altered by someone without permission. Software integrity ensures that the programs in the system have not been altered, whether by an error, a malicious user, or a virus. To a large extent, confidentiality is about unauthorized reading of data and programs while integrity is concerned with unauthorized writing.

Cryptographic algorithms form a set of primitives that can be used as building blocks to construct security mechanisms that serve specific objectives. For example, network security protocols, such as IPSec and, SSL combine these primitives to achieve authentication between communicating entities, and ensure the confidentiality and integrity of communicated data. These mechanism are referred to as *"functional security mechanisms,"* since they only specify what functions are to be performed, irrespective of how these functions are implemented. For example, the specification of a security protocol is usually independent of whether the encryption algorithms are implemented in software running on an ad hoc processor, or using custom hardware units, and whether the memory used to store intermediate data during these computations is on the same chip as a computing unit or on a separate chip. The separation of concerns between functional security mechanisms and their implementation has enabled rigorous theoretical analysis and design of cryptosystems and security protocols. In the process, however, various assumptions are made about the implementation of functional security mechanisms. For example, it is typically assumed that the implementation of cryptographic algorithms are ideal "black boxes," whose internals can neither be observed nor interfered with by any malicious entity. Under these assumptions, the level of security is widely quantified in terms of mathematical properties of the cryptographic algorithms and their key lengths.

In practice, functional security mechanisms alone are far from being complete security solutions. It is unrealistic to assume that attackers will directly attempt to take on the computational complexity of breaking the cryptographic primitives employed in security mechanisms. Rather, they will look for weaknesses in the implementation and deployment of functional security mechanisms and their cryptographic algorithms. These weaknesses can allow attackers to bypass completely, or significantly weaken, the theoretical strengths of security solutions. Such implementation vulnerabilities abound in current node hardware due to:

(1) *Untrustworthiness of operational environment* It is expected that ad hoc nodes guarantee secure operation even when under the possession of owners who cannot be trusted. It is a lot easier to design a system if we can rely on inbuilt physical security of the device, or assume that parts of the system cannot be physically accessed by malicious operators. However, ANs are sometimes required to work under complex trust relationship, where one party wants to put a secure device in the hands of another, with the assurance that the second party cannot modify the internals of the secure device. For example, being smaller in size, ad hoc nodes are more prone to loss or theft and as a consequence can remain in the hands of malicious users for extended periods of time.

(2) *Networking* Ad hoc nodes may have inbuilt networking interfaces that expose them to many sources of attack on the security mechanisms and that can be launched by remotely located entities. Wireless interfacing makes these nodes inherently vulnerable and if there is connectivity to the Internet, then the vulnerability is further compounded.

(3) *Downloaded software* The software that ad hoc nodes use may be third-party software purchased from a vendor, downloaded from the Internet. Such software pieces may contain viruses, worms, and Trojan horses, which are preferred instruments of malicious users for launching attacks.

(4) *Commercial off-the-shelf components* The ad hoc node's hardware is typically assembled using components from multiple sources. The responsibility for ensuring system security typically falls upon the manufacturer of the end product that is sold, or upon the entity that provides services based on the end product. Furthermore, even if each part of a system is secure in itself, it is known that composition of parts may expose new vulnerabilities [4].

Designing systems that are absolutely *tamper-proof* is often not possible, mainly due to two reasons: (i) very high costs incurred in assembling a device that can withstand a large number of often unknown attacks [5], and (ii) constant improvements in technology provide hackers with new tools to increase their reach and capabilities to launch sophisticated attacks. As a consequence, the approach that is taken in industry is to design systems that are *tamper-resistant*.

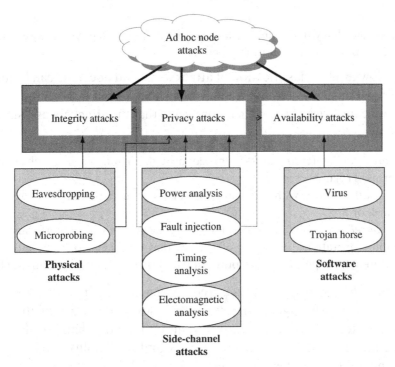

Figure 3.1 Classification of attacks on an ad hoc node

3.7.2 *Attacks on secure ad hoc nodes*

The previous section explained that to achieve high levels of security requires strong functional security mechanisms that are embodied in tamper-resistant implementations. The design of tamper-resistant implementations requires a strong awareness of the potential implementation weaknesses that can become security flaws, and careful consideration of security during all aspects of the architecture, hardware and software design processes. In this section we describe major attack techniques that can threaten the security of an ad hoc node. Countermeasures for the prevention and detection of attacks along with the recovery schemes that can be used after an attack are described in [3].

Figure 3.1 shows a high-level classification of attacks on an ad hoc node. At the top level, attackers are classified into three main categories based on their functional objectives.

(1) *Integrity attacks* The purpose behind these attacks is to try to change the data or the code on the ad hoc node.
(2) *Privacy attacks* The idea behind these attacks is to gain access to sensitive information stored, communicated, or manipulated within an ad hoc node.

(3) *Availability attacks* These attacks belong to the category of denial of service attacks, which try to disrupt the normal functioning of the system by exhausting the resources needed for normal operation.

A rather low level of classification of attacks on an ad hoc node can be arrived at by considering the means that an attacker uses to initiate the attacks. The three main categories of the vehicles or agents that the attackers use, shown in Fig. 3.1, are:

> *Software attacks* refer to attacks launched through software agents, such as viruses, worms, Trojan horses, etc.
>
> *Physical attacks* deal with physical abuse of the system at some level, e.g., chip, board, or box.
>
> *Side-channel attacks* are based on monitoring the system while it performs cryptographic operations: examples of entities that are likely to be monitored are execution time, power consumption, and node behavior in the presence of faults.

The agents that launch the attacks, could, again, be active or passive (see Fig. 3.2). The passive agents simply observe some properties of the system without interfering with it. An active agent, on the other hand, will interfere with the operation of the system. Integrity and availability attacks require interference with the system in some manner, and hence can be launched only through active agents.

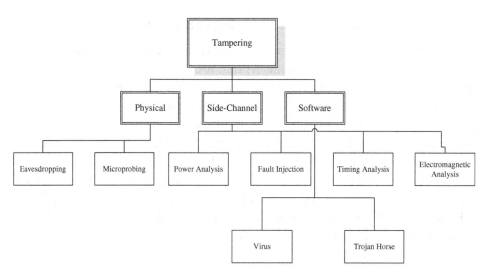

Figure 3.2 Tampering attacks on an ad hoc node

While the classification given above helps in understanding different types of attacks on a node, one needs to recognize that attackers rarely launch a particular attack in its pure form. Instead they use combinations of different attacks to realize their goals. For example, physical attacks may be used as a precursor to side-channel attacks, which means removing the packaging of the chip first and then observing the activities of the bus on the chip.

Software attacks

Software attacks are a major threat to an ad hoc node: they are launched using malicious agents such as a worm or a virus and can affect the system security attributes such as integrity, privacy and availability. This type of attack requires a simple and inexpensive infrastructure. Malicious agents typically look for the weaknesses in the system architecture which arise due to short-comings in the software. These shortcomings are known as *vulnerabilities* or *exposures*. A software *vulnerability* allows the attacker to gain direct access to the end system, while an *exposure* provides an entry point that can be exploited to gain access. A buffer overflow problem [6] is a common loophole in operating systems and application software and is an excellent example of a software attack. The buffer overflow effects include overwriting stack memory, heaps, and the function pointers. The attacker can use buffer over-flows to overwrite program addresses stored nearby. This may allow the attacker to transfer control to malicious code, which, when executed, can have adverse effects.

Physical and side-channel attacks

Various physical and side-channel attacks can be launched against an ad hoc node. Some of the prominent attacks are discussed below:

Physical attacks The first step in such attacks is de-packaging. De-packaging involves removal of the chip package by dissolving the resin covering the silicon using fuming acid. The next step involves layout reconstruction using a systematic combination of microscopy and invasive removal of covering layers. During layout reconstruction, the internal structure of the chip can be inferred at various granularities. While higher-level architectural structures within the chip, such as data and address buses, memory and processor boundaries, etc., can be extracted with little effort, detailed views of lower-level structures such as the instruction decoder and ALU in a processor, ROM cells, etc., can also be obtained. Finally, techniques such as manual microprobing or electron-beam microscopy are typically used to observe the values on the buses and the interfaces of the components in a de-packaged chip. Physical attacks at the chip level, however, are relatively hard because of their expensive infrastructure requirements relative to other attacks. But once done, they provide enough information for launching non-invasive attacks. For example, the knowledge of ROM contents, such as cryptographic routines and control data, can provide an attacker with information that can assist in the design of suitable non-invasive attack.

Power analysis attack The power consumption of any hardware circuit, e.g. a processor running cryptographic software, is a function of the switching activity at the wires inside it. Since the switching activity (and hence power consumption) is data dependent, it is not surprising that the key used in a cryptographic algorithm can be inferred from the power consumption statistics gathered over a wide range of input data. Power analysis attacks have been shown to be very effective in breaking small embedded systems [7]. Power analysis attacks are categorized into two main classes: simple power analysis (SPA) and differential power analysis (DPA) attacks.

Simple power analysis attacks rely on the observation that, in some systems, the power profile of cryptographic computations can be directly used to reveal cryptographic information such as the algorithm used, the cryptographic operations being performed, etc.; they require a reasonably high resolution to reveal the cryptographic key. Differential power analysis attacks [7] employ statistical analysis to infer the cryptographic key from power consumption data. This attack uses the differences between traces to overcome the disadvantages of measurement errors and the noise associated with SPA. Differential power analysis has been shown to be effective in extracting symmetric keys as well as public keys from several embedded systems.

Timing analysis Timing attacks [8] are based on exploiting the fact that execution times of cryptographic computations are data dependent and, hence, can be used to infer cryptographic keys. Execution time can vary, depending upon the implementation of the algorithm as well as the architecture of the hardware. However, by collecting the execution time statistics of the cryptographic algorithm and subsequently analyzing the wide range of data it is possible to deal with the differences in implementations to break the key.

Fault injection attacks Fault injection attacks rely on varying the external parameters and environmental conditions of a system, such as the supply voltage, clock, temperature, radiation, etc., to induce faults in its components. The injected faults can be transient or permanent and can compromise the security of a system in several ways such as:

(1) *Availability attacks* By injecting the hardware faults in a system, its normal operation can be disrupted and this may lead to a denial-of-service type of attack. For example, if a bus is made unavailable by injecting a fault (e.g., change in voltage) no communication between different elements of the system can take place.

(2) *Integrity attacks* The attacks are used to corrupt the code or the data by modifying the contents stored in the memory.

(3) *Privacy attacks* It has been shown that the RSA modulus [9] can be factored very easily if faults can be introduced to affect the outputs of one of the exponentiations being performed, thus revealing the cryptographic keys.

(4) *Pre-cursor attacks* Fault injection techniques are also useful as a precursor to software attacks. For example, it has been shown that simple memory faults induced by heat can be exploited by an untrusted program running on a processor to assume complete control of its execution environment [10].

Electromagnetic analysis attacks These attacks are based on the observation that the electromagnetic radiation emitted from a video display unit can be used to reconstruct its screen contents [11]. The basic premise of this attack is that it attempts to measure the electromagnetic radiation emitted by a device to reveal sensitive information [12].

3.8 Further reading

This chapter discussed various attacks on ad hoc nodes and presented their classification. It also presented several attacks, such as integrity attacks, privacy attacks and availability attacks. The chapter then presented yet another classification of attacks on an ad hoc node by considering the means that the attackers use to launch attacks. The primary means that the attackers used were software attacks, physical attacks, and the side-channel attacks that can lead to tampering of an ad hoc node. There are several solutions that have been proposed to make the design of an ad hoc node more tamper resistant, which are discussed in references [13] and [14]. The treatment of various topics related with the threats and attacks has been rather brief. Readers seeking additional information should see the references cited in the chapter.

3.9 References

[1] J. Newsome, E. Shi, D. Song, and A. Perrig, "The Sybil attack in sensor networks: analysis and defenses," *3rd Int. Symposium on Information Processing in Sensor Networks*, 2003, pp. 171–179.

[2] C. Karlof and D. Wagner, "Secure routing in wireless sensor networks: attacks and counter measures," in *First IEEE Int. Workshop on Sensor Network Protocols and Applications*, 2003, pp. 113–127.

[3] S. Ravi, A. Raghunathan, and S. Chakradhar, "Embedding security in wireless embedded systems," in *Proc. Int. Conf. VLSI Design*, 2003, pp. 269–270.

[4] J. Kelsey, B. Schneier, and D. Wagner, "Protocol interactions and the chosen protocol attack," in *Proc. Int. Workshop on Security Protocols*, 1997, pp. 91–104.

[5] R. Anderson and M. Kuhn, "Tamper resistance – a cautionary note", in *Proc. Second USENIX Workshop on Electronic Commerce*, Oakland, CA, 1996, pp. 1–11.

[6] E. Chien and P. Szor, "*Blended Attacks Exploits, Vulnerabilities and Buffer-Overflow Techniques in Computer Viruses,*" Symantec white paper, www.symantec.com/avcenter/reference/blended.attacks.pdf, 2002.

[7] P. Kocher, J. Jaffe, and B. Jun, "*Introduction to Differential Power Analysis and Related Attacks*," www.cryptography.com/resources/whitepapers/, 1995.

[8] P. C. Kocher, "Timing attacks on implementations of Diffie–Hellman, RSA, DSS, and other systems," *Advances in Cryptology – CRYPTO'96, Springer-Verlag Lecture Notes in Computer Science*, vol. 1109, 1996, pp. 104–113.

[9] D. Boneh, R. DeMillo, and R. Lipton, "On the importance of checking cryptographic protocols for faults," in *Proc. Eurocrypt'97*, 1997, pp. 37–51.

[10] S. Govindavajhala and A. W. Appel, "Using memory errors to attack a virtual machine," in *Proc. IEEE Symposium on Security and Privacy*, 2003, pp. 154–165.

[11] W. Van Eck, "Electromagnetic radiation from video display units: an eavesdropping risk?," *Comput. Secur.*, vol. 4, no. 4, 1985, pp. 269–286.

[12] J. J. Quisquater and D. Samyde, "ElectroMagneticAnalysis (EMA): measures and counter-measures for smart cards," *Lecture Notes Comp. Sci. (Smart Card Programming and Security)*, vol. 2140, 2001, pp. 200–210.

[13] S. Ravi, A. Raghunathan, P. Kocher, and S. Hattangady, "Security in embedded systems: design challenges," *ACM Transactions Embedded Computer System*, vol. 3, no. 3, 2004, pp. 461–491.

[14] S. Ravi, A. Raghunathan, and S. Chakradhar, "Tamper resistance mechanisms for secure, embedded systems," *17th International Conference on VLSI Design*, 2004, p. 605.

4

Trust management

The previous chapters discussed mandatory security requirements, which include confidentiality, authentication, integrity, and non-repudiation. These, in turn, require some form of cryptography, certificates, and signatures. Some other security-related mechanisms include user authentication, explicit transaction authorization, end-to-end encryption, accepted log-on security (biometrics) instead of separate personal identification numbers (PINs) and passwords, intrusion detection, access control, logging, and audit trail. In this chapter, I present some of the security schemes that govern trust among the communicating entities. Governance of the trust can be based on principles and practices of key management in distributed networks or other means such as authentication. Additionally, this chapter discusses several well known methods that are related to key management and authentication.

4.1 The resurrecting duckling

The resurrecting duckling security model [1] has been developed to solve the secure transient association problem. An example of this would be when a person buying a remote control would not want any other person to be able to use another remote control bought at the same shop to work at his place, but then the remote control has to work for some other person who might buy it from the first owner. Like a duckling, who considers the first moving object it sees to be its mother, in the same way a device would recognize the first entity that sends it a secret key as its owner. When necessary, the owner could later clear the imprinting and let the device change its owner. The imprinting – sharing the key – would be done in a physical contact. In the case of several owners with different access rights, the imprinting could be done several times with different keys. In this manner, it could be possible to create a hierarchy between the owners, or prioritize the service requests. Tamper resistance, or tamper evidence, may protect against physical threats of the nodes. In this scheme, the uniqueness of a master is emphasized. A slave has two exclusive states: imprinted and imprintable. The master controls the slave and they are bound together with a shared secret that is originally transferred from master to slave over a non-wireless, confidential, and integrated channel. The slave is imprinted and made imprintable by the master. The slave becomes imprintable as a consequence of conclusion of a transaction or by an order of the master. The original "resurrecting duckling" security policy was extended to cover a peer-to-peer interaction [2]. In the extension, a master can be a human being in addition to devices. A human master can imprint the slave device by using a personal identification number (PIN). A master can also be a sensor node, which is part of a network that is physically monitoring an event of interest in an area. A sensor node has a battery, solar cell, sensing mechanisms, such as sensors, and some digital computing hardware, besides an active transmitter and receiver. Another feature of this extended scheme is that the master does not have to be unique. The slave can also be imprinted by another master that has a credential which is valid for that slave at that moment. The slave has a principal master, but it can also receive some kinds of order from other masters. In this extension of the resurrecting duckling model, the master can upload a new policy in slave mode, too.

The next section deals with key management, which involves services like trust models, cryptosystems, key creation, key storage, and key distribution, each of which has been touched upon as part of security needs.

4.2 Key management

This section discusses the various schemes that have been proposed to ensure a secure key management function within an ad hoc network. To be able to protect nodes against eavesdropping by using encryption, it is necessary that the nodes must have made a mutual agreement on a shared secret key or have exchanged public keys. For rapidly changing ad hoc networks, the exchange of encryption keys may have to be addressed on demand, thus without assumptions about a priori negotiated secrets. In somewhat static environments, the keys may be mutually agreed upon proactively or even configured manually.

4.2.1 A distributed asynchronous key management service

Tactical operations of the military are still one of the key applications of ad hoc networks and the military operations by nature are security sensitive. When mobile nodes are used in hostile environments, such as the battlefield, it is quite possible that a set of nodes could be captured and compromised. The presence of compromised nodes in a network allows for the possibility of launching internal attacks. For this particular reason, it is preferable not to have centralized nodes that act as certifying authorities. Distributed architectural solutions will be more suited and will enhance the security of the ad hoc networks.

As has been said before, no single node of an ad hoc network can be trustworthy because of unreliable physical layers, so it makes sense to develop algorithms that distribute the trust among several nodes. One such approach, described below, relies on the consensus of at least $(t + 1)$ out of n nodes, where $n \geq 3t + 1$.

In this discussion regarding distribution of trust, my focus is on securing the routing information in an ad hoc network. Ad hoc routing protocols face two types of threats. The first threat is from external attackers, who inject erroneous routing information, thereby distorting the routing behavior of the network. The second threat is more severe, as this is created by compromised nodes, which can create incorrect routing information that can be difficult to detect because the routing information might have been signed by a valid signature. Protection against the first kind of attack could be provided by the use of cryptographic schemes such as digital signatures but such schemes do not work for the second type of attacks, which are launched by compromised nodes. As discussed above, the detection of compromised nodes through routing information is difficult because it is hard to distinguish whether the invalid information is created by compromised nodes or has become invalid due to the topology change.

The routing of packets in a network consisting of a few compromised nodes is accomplished by bypassing these nodes by the routing protocol and discovering alternate routes. One of the security schemes that protects the routing information as well as the data traffic is based on public key cryptography. In a public key cryptography, each node has a public/private key pair. A public key is distributed to every node that needs it, while private keys are known to individual nodes and are not shared. Public key cryptography requires the existence of a secure entity known as a certification authority (CA) for key management. The CA has a public/private key pair, whose public key is known to every node. The CA issues certificates that bind a public key to a node. As I have stated earlier that, for ad hoc networks, it is not a good idea to have a centralized CA, the scheme discussed below advocates a distributed CA architecture in which the certification authority is formed by a set of nodes that jointly manage the key management responsibility.

A distributed key management algorithm in which the private key of a trusted service is divided and distributed to n servers is shown in Fig. 4.1.

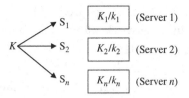

Figure 4.1 Key management service K/k

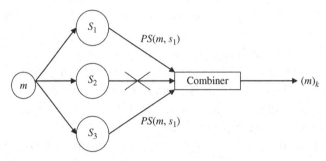

Figure 4.2 Key management and threshold signature

The service, as a whole, has a public/private key pair K/k. The public key K is known to all nodes in the network, whereas the private key k is divided into n shares $S_1, S_2, \ldots, S_n$, one share to each server. Each server i also has a public/private key pair K_i/k_i and knows the public keys of all nodes.

To create a signature with the private key, at least k out of the n servers need to combine their knowledge. Combining the shares would not reveal the actual private key. The correctness of the signature would, as usual, be verifiable with the public key of the service. This method is called *threshold cryptography*: an (n, k) threshold cryptography scheme allows n parties to share the ability to perform a cryptographic operation (e.g., creating a digital signature). This is shown in Fig. 4.2 for three servers where K/k is the public/private key pair of the service. Using a (3, 2) threshold cryptography scheme, each server i gets a share s_i of the private key k. For a message m, server i can generate a partial signature $PS(m, s_i)$ using its share s_i. Correct servers S_1 and S_3 both generate partial signatures and forward the signatures to a combiner c. Even though server S_2 fails to submit a partial signature, c is able to generate the signature $(m)_k$ of m signed by service private key k.

Any k parties can perform the operation jointly, whereas it is infeasible for, at most, $k - 1$ parties to do so. If we suppose that, at most, $k - 1$ servers can be compromised at a time, a false signature cannot be created. The key management service also employs share refreshing and is scalable to changes in the number of servers. Periodical share refreshing creates new shares of the private key, so that an adversary cannot collect information about k shares over time. In effect, the scheme makes use of redundancies in the network topology to provide reliable key management and implements a distributed model due to the lack of a central authority. The key idea of this algorithm is to use key sharing with the assumption that the ratio between nodes compromised to total nodes is bounded. If the upper limit on the number of compromised server nodes can be set to $t > 1$, at least $n \geq (3t + 1)$ nodes are needed to enable the scheme. The proposed architecture does require that the underlying routing protocol manages multiple routes.

4.2.2 A password-authenticated key exchange protocol

A generic protocol for multi-party password-authenticated key exchange has been proposed, based on a pioneering piece of work in which two parties A and B share a weak secret P [3]. The underlying protocol's goal is to agree to a strong session key K in spite of weak P, in such a way that an attacker watching the traffic will not be able to learn K or mount an attack on P. A has a random key pair (E_A, D_A) for encryption and decryption respectively.

During the protocol operation, A generates two random strings: challenge$_A$, and S_A. B generates three random strings: R, challenge$_B$ and S_B. The protocol operates as follows. In the first step, A sends his or her identifier and an encrypted weak secret, P, to B. In the second step, B extracts the encrypted key for A (called E_A), and generates a random string R. Then R is encrypted with E_A and returned back to A. In the third step, R is extracted and random strings challenge$_A$ and S_A are generated. They are encrypted by R and sent back to B. In the fourth step, B extracts challenge$_A$ and computes a public function h(challenge$_A$). B's own random strings, challenge$_B$ and S_B, that are generated and encrypted with h(challenge$_A$) using R as a key are sent back to A. A extracts the three quantities from this message and verifies that the first quantity is indeed h(challenge$_A$). At this point A can be convinced that B can extract challenge$_A$. This is possible only if B knows the weak secret, P, that was used. A then computes h(challenge$_B$), encrypts it using R as the key, and returns the result to B in step 5. B decrypts the message and verifies that the message received is indeed h(challenge$_B$). B can now be certain that A was able to extract challenge$_B$ from the message in step 4, which in turn implies that A knows P and was able to extract R correctly from the message in step 2. At this point, each player will compute the session key as $K = f(S_A, S_B)$. The protocol can be extended to a multi-party scenario by electing a leader. In that case, S functions are also used to generate a strong session key K. Other parties'

(1)	A → B :	A, $P(E_A)$
(2)	B → A :	$P(E_A(R))$
(3)	A → B :	R (challenge$_A$, S_A)
(4)	B → A :	R (h(challenge$_A$), challenge$_B$, S_B)
(5)	A → B :	R (h(challenge$_B$))

Figure 4.3 Protocol for password-authenticated key exchange

knowledge of K has to be confirmed. Figure 4.3 provides an overview of the underlying scheme.

The password-based authentication protocol is derived from the so-called encrypted key exchange (EKE) protocol. In EKE, two participants, who share a secret, create together a session key. The secret, or password, can be weak. Nevertheless, anyone who does not know the password cannot successfully participate in the protocol. Finally, EKE provides perfect forward secrecy: even if an attacker later finds out the password, he or she cannot find out the previous session keys. Hence, the messages of the past sessions remain secret. In the group version of EKE, all the participants contribute to the session key. This ensures that the resulting key is not selected from too small a key space, even if some participants try to do that. An attacker, who tries to participate in the protocol and sends some random messages, cannot prevent the construction of the key. As in the original EKE, only the participants who know the original password learn the resulting session key. The secure connections between the participants are created from a manually exchanged password. Hence, no support infrastructure is needed. The key agreement is of significant importance when a secure transient association is to be guaranteed. This will happen best when the key management is done locally. To implement location-based key agreement successfully, a set of labels to map the location as well as an identity-based mechanism for key agreement is needed.

4.2.3 A progressive trust negotiation scheme – NTM

A scheme for progressive trust negotiation in ad hoc networks has been proposed, which builds trust, along with a dynamic key agreement scheme to protect the negotiation [4]. The NTRG trust model (NTM) is sub-divided into two main components, namely, the peer-to-peer component and the remote component. The peer-to-peer component deals with securing communication with neighbors. The remote component has the dual responsibility of carrying out trust negotiation and establishing secure end-to-end communication.

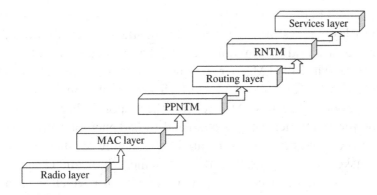

Figure 4.4 Layer structure

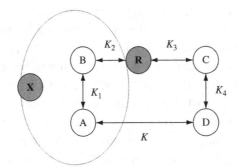

Figure 4.5 Key formation in NTM

The security system is divided into two distinct layers: the peer-to-peer NTM (PPNTM) layer and the remote NTM (RNTM) layer. The PPNTM layer is situated below the routing layer, as its primary goal is to secure communication between neighbors (i.e., nodes in the radio range). The RNTM layer provides end-to-end encryption and, so, is located just above the routing layer, as shown in Fig. 4.4.

The threat of eavesdropping by an external attacker, say X, which is within the listening range of two nodes A and B, is mitigated by the PPNTM layer. Since the symmetric encryption keys are to be negotiated between the neighbors using a station-to-station (STS) protocol, it is impossible for a node to eavesdrop on communication without authenticating itself. The end-to-end key negotiated by the RNTM layer protects against an internal attacker, R. This key formation in the NTM scheme is shown in Fig. 4.5, where K_1, K_2, K_3, and K_4 are peer-to-peer keys and K is the end-to-end key between A and D.

Each node has to have at least one network address certificate that entitles it to use certain network addresses and participate in packet relaying. This certificate is used in the PPNTM layer's STS key exchange for authentication. The RNTM layer relies on the identity certificate, so that a user can move between nodes while still maintaining trust. The certificates have to be signed by a third party. The PPNTM layer negotiates a symmetric key with neighbors. The remote certificate involved in the STS key formation is authenticated using three models. The RNTM layer has the responsibility of carrying out trust negotiations and negotiation of an end-to-end encryption key. The trust negotiation is carried out by incrementally exchanging certificates. The certificates are asked for by using the attribute name/value pair. Usually a node trying to access some services on the remote node, for which it hasn't been cleared, triggers the trust negotiation. The trust negotiation can also be explicitly triggered by the RNTM layer. The different models for finding trust in a certificate are necessitated by the absence of an online trusted third party. The simplistic first model assumes that the node is primed for local use and has all the certificate revocation lists (CRLs) updated. Any certificate issued by an unknown certificate issuer cannot be verified and will be referred to the user. The second model is a probabilistic model. Each of the trusted certificate issuers has a trust value of 1 associated with it. There is a distrust value, which is subtracted from the trust value of the CRL if the scheduled update of the CRL is missed. Then the trust negotiation takes place with a default trust value and it should be exceeded for negotiation to succeed. The third and last model assigns weights to the certificates.

4.2.4 Minimal public-key-based authentication

Sufatrio and Lam have introduced a lightweight and scalable authentication protocol called minimal public-key-based authentication (Min-PKA) for a mobile IP that does not require any changes to the protocol. Its main purpose is to secure the registration process. It makes use of AAA (authentication–authorization–accounting) server nodes (AAAH for AAA home agent and AAAF for AAA foreign agent). The authors criticize Jacobs' approach [5], which uses only public-key cryptography, since it assumes that the mobile nodes (MN) can perform the heavy computations related to the security operations. In contrast, the Min-PKA proposal uses two different approaches, secret-key-based and public-key-based, of which the former requires manual configuration. Since such an approach may offer substantial optimizations in some routing scenarios, they suggest the use of public-key cryptography to be applied in the inter-domain authentication. The mobile node and AAAH can, however, use shared secrets between the home agent (HA) and the mobile node, since the nodes have a security association.

Their approach introduces three services:

(1) Authentication services provide digital signatures and message authentication codes (MAC) between the MN and the AAAH to protect the routing traffic. The services rely on the correct actions performed by the AAAH in the indirect MN–AAAF communication.
(2) Integrity services rely on the authentication services to assure integrity when the authenticity is confirmed. Foreign agent (FA) discoveries form a problem, since the FA and the MN may have no security associations. This problem is, however, solved by putting the advertizements into the registration requests from which a MAC can be calculated and which can thus then be authenticated.
(3) Anti-replay protection services guarantee the freshness and authenticity of the registrations. The mechanism uses nonces, which are time stamps, to achieve the goals but has a flaw, since adversaries can fool the AAAFs to sign arbitrary data. Nonces in the message to be signed somewhat reduce the severity of the problem but do not completely remove it.

4.2.5 Non-disclosure method

Fasbender *et al.* have introduced the non-disclosure method (NDM) [6], a solution to the confidentiality-of-location problem, wherein the current location of a mobile node can easily be retrieved by just looking at the address headers of the exchanged packets and particular registration requests can then be used to generate location profiles. In this approach, every security agent (SA) node has a public/private key pair. When a sender A wants to send a message M to a receiver B, the message is forwarded to the destination by using a route $(A, SA_1, SA_2, \ldots, SA_n, B)$ as defined by the intermediate security agents from SA_1 to SA_2. The route is constructed by performing n encryptions E_SA_i with the public keys of the intermediate nodes: encrypted message $M' = E\_SA_1$ $(SA_2, E\_SA_2 (SA_3, \ldots (SA_n, E\_SA_n (B,M))))$. When the sender A sends the encrypted message M', the first security agent SA_1 decrypts the message, thus finding only the location of the next hop in the route SA_2 and so on. Thus, the security agents see only the location information (addresses) of the next and previous security agents. In addition, the nodes cannot determine where they actually are located in the route and who the receiver B is. In this approach, the last intermediate node SA_n would know the location and identity of the receiver B, but not M, if it can be assumed that the sender A can encrypt the message with B's public key also. The method can be applied to protect any other vulnerable header information rather than just the location of the nodes. In the NDM approach, the location information as well as the actual message is hidden from the intermediate nodes (SAs). This approach, however, has a problem with respect to MANET networking: the sender must know all the public keys K_SA_i and the identities of the security agents in the route to be able to construct the route. Moreover, the intermediate compromised nodes (or outsiders) can inspect the sizes of the sent packets and try to determine the length of the route. This problem can be mitigated by allowing the SAs to use padding mechanisms with random data to hide the actual length of the payload.

4.2.6 Securing ad hoc Jini-based services

Jini is an open software architecture that enables developers to create network-centric services that are highly adaptive to change and a security model based on the usage of public keys has been proposed for securing ad hoc Jini-based services [7]. Unlike most of the other service discovery methods, Jini uses a distributed model that is built on allowing code to be moved between entities in the ad hoc network. Downloaded code is a security problem in itself. Hence, security is an important part of the system design of Jini ad hoc services, and the authors address it by relying on decentralized authorization. The problem of trusting public keys is separated from the usage of trusted keys. Given that all of the nodes in the ad hoc network have public-key pairs, and that all of the nodes consider the public keys of others good for creating secure connections within the network, any of the public-key-based authentication schemes can be used. It is possible to use transport-layer security protocol (TLS) or IPSec protection protocols to secure the actual service. The authors' solution allows the service provider to design a communication security solution that is adapted to the service and uses any cryptographic algorithm. Unlike other standard approaches, their approach does not assume that the client and the server share a necessarily large set of different symmetric key encryption or MAC algorithms. The authors have suggested a trust distribution protocol that minimizes the number of manual interactions needed when setting up the necessary trust relations, as shown in Fig. 4.6. A server that wants to offer a secure communication service has a proxy, which contains the necessary algorithms for authenticated key exchange with the server, and also algorithms to encrypt and protect the exchanged data in the client–server interaction. The server digitally signs the proxy with its private key, which ensures that the client can verify the signature. The server packs the signed code together with the signature (and possibly its public key). When a client finds the service, it downloads a proxy corresponding to the service, together with the signature and possibly some included certificates. The client can verify the signature if it has a trusted public key that corresponds to the signature or if the client trusts

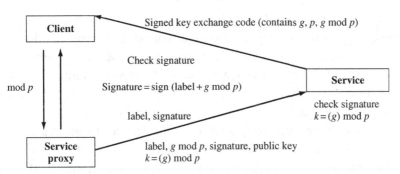

Figure 4.6 A basic proxy distribution protocol with the proxy using Diffie–Hellman key agreement

some of the public keys contained in the included certificates. The proxy performs authenticated key exchange with the origin server using a standard authentication and key exchange protocol. On successful authentication, the proxy sets up a secure communication link with the server.

Furthermore, the authors have described a "minimal" pre-configuration solution for securing the communication of an ad hoc service. The technique allows establishment of authenticated connections without allowing undue access to the client's private key. The example implementation uses Java and Jini, but the same principles can be used with other service discovery techniques.

4.2.7 A robust ubiquitous security scheme

A scheme for providing ubiquitous security services for mobile hosts has been proposed, which scales to network size, and is robust against break-ins [8]. The scheme distributes the certification authority functions through a threshold secret-sharing mechanism, in which each entity holds a secret share and multiple entities in a local neighborhood jointly provide complete services. Localized certification schemes are employed to enable ubiquitous services. Secret shares are updated to enhance robustness against break-ins. Threshold secret sharing and secret share updates are used to enable intrusion tolerance. No single entity in the network knows or holds the complete system secret (e.g., a certification authority's signing key). Instead, each entity holds only a secret share of the certification authority's signing key. Multiple entities, say K, in a one-hop network locality jointly provide complete security services, as if they were provided by a single and omnipresent certification authority. The system security is not compromised, as long as there are fewer than K collaborative intruders in each adversary group. To resist intrusions on a long-term basis further, the secret shares for all entities are periodically updated. A certificate-based approach based on the public key infrastructure (PKI) is employed. Any two communicating entities may establish a temporary trust relationship via unforgeable, renewable, and globally verifiable certificates carried by each of the entities. Security functions such as confidentiality, data integrity, authentication, and non-repudiation can be readily provided via valid certificates that are usually issued by a globally trusted certification server.

New schemes have been proposed to realize the certificate-related security services and accommodate the unique characteristics of ad hoc wireless networks. They provide ubiquitous services for mobile entities by distributing the certification authority's functionality to each local neighborhood. A coalition of K neighbors can serve as the CA and jointly provide certification services for a requesting mobile entity. The fully localized and universally available features of their design enable service ubiquity for mobile users. A novel self-initialization protocol is proposed to handle dynamic node membership (i.e., joins and leaves) and secret share updates. Each node can be (re)initialized by K neighbors. Once initialized, a node is qualified to be a coalition member to serve its neighborhood. Security services are effectively provided in the presence of mobility, wireless channel errors, network partitioning, and node failures.

4.2.8 Robust membership management scheme

A robust membership management scheme for ad hoc groups based on public key cryptography and on the use of signed certificates has been described, in which the members are represented by their public signature keys and each group has a public signature key to represent the group as a whole [9]. Certificates signed by the group key are used to indicate the membership of the nodes. In groups, the owner of the group key is the only member who can let new members join the group. To increase the robustness of the membership management, the authority of the leader must be distributed to several members. This is done by letting the original leader (i.e., the group-key owner) delegate the leadership to other members. It can authorize other keys to act as equivalent leaders by issuing leader certificates. To prove membership in the group, a member that has been certified directly by the group key needs its own private key and the membership certificate. A member that has been certified by another leader needs all the certificates in the path from the group key to its member key. Hence, when a leader certifies other leaders or members, it must pass along all the certificates that prove its own status as a leader in the group. In this way, a chain of certificates is formed from the group key to each member key.

Reconstitution of the group is a secure and often a recommendable way to continue with the trusted members only. The members of a group also need some instant mechanisms for canceling the membership of a single key without sacrificing membership of the other, still trusted members. Unfortunately, canceling a membership that has already been granted is not easy. The membership certificates may be created, stored, and verified concurrently at different parts of the system. There are two ways of getting rid of untrusted members: membership expiration (the membership certificates may have a validity period that is decided by the issuer) and membership revocation (members should be revoked only when there is a reason to suspect that the private key has fallen into the hands of an adversary and information about the revocation must be propagated to all the parts of the system where relevant certificates may be used).

Increasing the robustness of the scheme, erasing the group key and issuing redundant certificates have been dealt with by the authors of the scheme as discussed below:

Increased robustness with erased group keys and redundant certificates One effect of the expiration or revocation of a leader key is that it not only causes the removal of that leader but it also affects every member below the removed leader in the tree structure formed by the certificates. The result is particularly dramatic if the membership of the group key itself is revoked. For if the group key becomes under suspicion and needs to be revoked, the whole group perishes. Revocation of the group key may be desired when one wants to replace the group key with a new one and reconstitute the entire group.

Erasing the group key A perfect way of protecting the private key against a compromise is to erase it. An erased key cannot be recovered or misused in any way. The certificates signed with the erased key continue to be valid and they can still be verified with the public key. In the group context, the newly generated private group key can be used to certify a few leaders and then erased. Several leaders should be certified with the group key so that if the membership of one of them must be revoked, the remaining leaders can still continue to administer the group. The certificates signed by a key cannot be refreshed after erasure of the key, so when these leader certificates are about to expire, the group needs to be reconstituted. This is shown in Fig. 4.7. In the group context, the newly generated private group key can be used to certify a few leaders and then erased. Several leaders should be certified with the group key so that if the membership of one of them must be revoked, the remaining leaders can still continue to administer the group. The group members should be informed that the group key is protected and cannot be compromised. In this way, they know that the group key will never be revoked. For example,

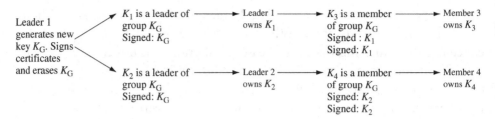

Figure 4.7 Erased group key

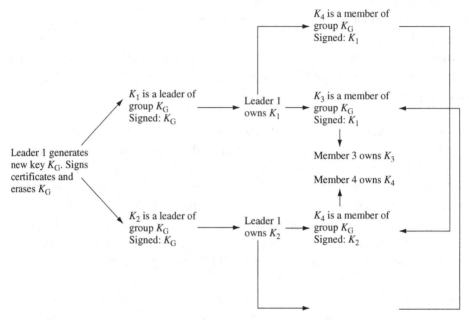

Figure 4.8 Redundant certificates

Leader 1 in Fig. 4.7 uses the group key to certify its own key and another key as leaders of the group. After signing the certificates, the group key is erased from Leader 1's memory.

Issuing redundant certificates Erasing the private group key removes a single point of failure, in the sense that there is no single key whose compromise would disable the entire group. However, large parts of the certificate tree can still be removed from the group by revoking one of the leaders certified by the group key. A member can alleviate this threat by obtaining multiple independent certificates. The leaders may also issue redundant certificates to each other. This is shown in Fig. 4.8. Even if Leader 2 loses its authority (its leader certificate expires or its membership is revoked), Member 4 still remains a member in the group. If the membership of either Member 3 or Member 4 is revoked, redundant certificates do not prevent or complicate the revocation in any way. In fact, by revoking the whole membership of an untrusted member, the leader need not be aware of the certificates that have been issued to the compromised member. This is why it has been chosen to revoke the memberships and not single membership certificates.

4.2.9 Scalable ubiquitous security scheme

Yet another design that supports ubiquitous security for mobile nodes, scales to network size, and is robust against adversary break-ins, is suggested by Luo and Lu [10]. In their design, they distribute the functionality of conventional security servers, specifically the authentication services, so that each individual node can potentially provide other nodes' certification services. Centralized management is minimized and the nodes in the network collaboratively self-secure themselves. A suite of fully distributed and localized protocols is proposed to facilitate practical deployment. These protocols also feature communication efficiency, to conserve the wireless channel bandwidth, and independency from both the underlying transport layer protocols and the network layer routing protocols. The focus is on the authentication service in ad hoc wireless networks. Their work is based on asymmetric cryptographic techniques, specifically the de facto standard RSA algorithms. Once the authenticated channels are established with proper access control between communicating parties, confidentiality, integrity, and non-repudiation can be further realized by following the typical Diffie–Hellman key exchange protocols.

These employ several techniques to achieve the design goals:

(1) Ubiquitous authentication service availability by taking a certificate-based approach. Any two communicating nodes establish a temporary trust relationship via globally verifiable certificates. With a scalable threshold sharing of the certificate-signing key, certification services, such as certificate issuing, renewal, and revocation, are distributed to each node in the network. No single node holds the complete certificate-signing key. Each node only possesses a share of it. While no single node has the power of providing full certification services, multiple nodes in a network locality can collaboratively provide such services that are the same as those that would be provided by an authority with a complete certificate signing key.

(2) By the distributed certification services, together with the further enhancement of a scalable proactive update mechanism, service robustness is ensured in the presence of short-term computation bounded adversaries.

(3) While the certification service distribution and periodical proactive update can be solved in theory using known cryptographic techniques such as threshold secret

sharing, threshold multi-signature, and proactive RSA, the approach focusses on scalable and practical solutions in large-scale ad hoc networks with dynamic node membership.

The proposed fully localized (typically within one-hop neighborhood) and distributed protocols further achieve communication efficiency and load balancing over the network to avoid network congestion [10]. Through the localized design, their communication protocols are immune from the unreliability of the underlying transport layer protocols and routing mechanisms in ad hoc wireless networks. Furthermore, their design has two additional features:

(1) Provable cryptographic security. Their proposed security algorithms are as secure as the underlying cryptographic primitives (e.g., RSA) by the simulatability arguments.
(2) Self-defensive, built-in detection mechanisms. While their design works with any intrusion detection algorithms and mechanisms that are of each individual node's choice, they apply the verifiable techniques as built-in mechanisms to detect adversaries that attack their security protocols.

4.3 Authentication

Authentication usually means that there is some way to ensure that the entity to which you are talking is who it claims to be. This is called authentication of the channel end point. Usually you also need to authenticate yourself to the service in order for the service to be sure that you are you, not someone else who is pretending to be you. This is the authentication of the message originator. The use of a password is not really a good choice, because passwords are typically short and easy to break. More secure methods include the use of public key cryptography, challenge-response schemes, symmetric encryption, etc.

4.3.1 The MANET authentication architecture

The MANET authentication architecture (MAA) proposed by Jacobs and Corson [11] places the emphasis on building a hierarchy of trust relationship to authenticate Internet MANET encapsulation protocol (IMEP) message security. The proposed scheme details the formats of messages, together with protocols that achieve authentication. The difficulty related with proactive schemes is that, first, cryptography is relatively computationally expensive on mobile hosts, where computational capability is comparatively restricted; second, since no central authority can be depended on, the authentication is more difficult to implement, and, third, it is only useful to prevent intruders from outside (external attack). If an internal node is compromised (internal attack), such schemes no longer work.

The MAA supports several authentication options, ranging from simple to complex. The IMEP authentication object is used to authenticate all IMEP messages between routers, which is accomplished by calculating an authenticator ("digital signature"). Also, MAA identifies the security context between a pair of MANET nodes. The certificate object, though optionally used, depending on the security context between corresponding MANET nodes, includes a copy or copies of certificates that bind system "distinguished names" to public keys using a digital signature. The trust hierarchy paths establish a logical chain between two certification authorities and establish trust relationships through intervening CAs. Certificate validation involves constructing a trust hierarchy path among the sender certificate, the certification authority that issued the sender certificate and the CA of the validating system. A trust hierarchy path must be established to verify authenticity and usability of certificates within IMEP. The receiver can develop trust in the public key of the sender's CA recursively, if the receiver has a certificate containing CA's public key signed by a superior CA whom it already trusts. Each certificate is processed in turn, starting with that signed using the input trusted public key. There is provision for certificate revocation lists (CRL) and certificates are checked against current CRLs from the issuing CA. A MANET node caches received certificates along with a value ("staleness value"). The node maintains a maximum ("staleness threshold") value of the certificate staleness tolerable before the node has to retrieve the appropriate CRL and verify that the certificate has not been revoked.

4.3.2 An end-to-end data authentication scheme

An end-to-end data authentication scheme for ad hoc networks that relies on mutual trust between nodes has been suggested, in which the basic strategy is to take advantage of the hierarchical structure that is implemented for routing purposes [12]. The scheme uses TCP at the transport layer and a hierarchical architecture at the IP layer so that the number of encryptions needed is minimized, thus reducing the computational overhead. Also, each node has to maintain keys for fewer nodes. The scheme makes use of a cluster-based network with a cluster-based routing protocol (CBRP) [13]. When a node joins the network, it is given a system public key and a system private key, as well as a cluster key. This cluster key is unique to the cluster. Each node has a table of cluster IDs and the corresponding head's public key. When a node joins a network for the first time, a strong authentication is done by sending a challenge and receiving a response. The system key pair is used for a mutual authentication between the joining node and an existing member of the network. When a node leaves a cluster and joins another cluster, the new cluster head treats this joining node as any other new node joining its cluster. A mutual authentication is performed between the moved node and its new cluster head using the system key pair. The new joining node gets the cluster ID, whereas the old cluster head purges the entry for the node, which moved out. When a node in one cluster wants to communicate with a node from another cluster, for complete confidentiality, the entire packet is encrypted with a session key. This session key is shared between only the two communicating parties and, thus, serves as authentication. For replay prevention, strong authentication may be performed for each packet, i.e., a series of challenges and responses back and forth.

The proposed algorithm involves exchange of a session key that is valid for just that particular TCP session after mutual authentication with the cluster heads acting as certification authorities. The heads generate a set of random prime numbers, which are first encrypted with each head's private key and then the cluster key. With each number, a time stamp is also encrypted for limited usage. The head then broadcasts them. These could serve as authentication tags for any of the cluster members (see Fig. 4.9). They are also encrypted with the session key for more protection. The receiver utilizes a check function to verify the origin and authenticity of the tags from the sender. The checksum field of the TCP header is also encrypted with the session key for more security. The check function has encrypted sequence numbers for which it is valid and, thus, packets cannot be replayed. On the flipside, the cluster head

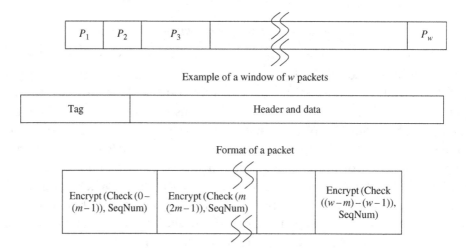

Example of a window of w packets

Tag	Header and data

Format of a packet

Encrypt (Check $(0-(m-1))$, SeqNum)	Encrypt (Check $(m-(2m-1))$, SeqNum)		Encrypt (Check $((w-m)-(w-1))$, SeqNum)

Figure 4.9 A packet containing checks for tags

needs to generate random prime number sets periodically and a session key needs to be generated for every session.

4.3.3 Authentication, authorization and accounting (AAA)

One way to deal with low physical security and availability constraints is the distribution of trust. Trust can be distributed among a collection of nodes [14]. Public key cryptography can be used for the *authentication* after having built a key management system. *Authorization* is also needed, because we do not want some malicious host to be able to wreak havoc inside the network. This can be prevented by keeping control of what hosts are allowed to do inside the ad hoc network. Authorization also needs some sort of distributed structure, because we cannot rely on one point of failure. In ad hoc networks, individual mobile hosts are providing a service to each other, which gives rise to *accounting*. For example, if some mobile node acts as a router in the network, providing connectivity between two nodes that are not within each other's range, then it would be reasonable to charge some money for this service. There exist no protocols to do the actual charging if that is needed. Because we cannot assume connectivity to some central server that takes care of the charging, there is a clear need for distributed charging protocols as well. Ad hoc networks and general AAA systems do not fit well together. The biggest problem is related to the varying nature of the network. There are no home domains or foreign domains, because the networks come and go on demand. This affects the AAA systems, because some of the basic building blocks of their architecture are missing in ad hoc networks. The basic problem here is that the general AAA model is a centralized trust model, whereas the ad hoc network structure is decentralized. There is a need for some other kinds of method to achieve the AAA functionality. One approach to provide authentication and authorization functionality in ad hoc networks could be to use trust management based approaches such as PolicyMaker [15] or Keynote [16], which are decentralized by nature and can provide the requested functionality in ad hoc networks quite easily. Also, other protocols such as SASL [17] or ISAKMP [18] and IKE [19] could be used to provide the authentication functionality. In ad hoc networks, we probably need decentralized models or some other approach to provide the AAA functionality.

4.4 Further reading

This chapter covered the topics related with the key management and authentication for wireless ad hoc networks. Key management and authentication of users are typically concerned with the broader area in network security called *trust management*. Authentication usually means that there is a possible way to ensure that the entity to which you are talking is what it claims to be. Very often, the network subscriber to the service also needs to be authenticated so that the service verifies the identity of the subscriber. Key management for wireless networks is still a very active area of research and several IEEE and ACM conferences, such as Infocom, Mobicom, and Mobihoc, as well as ICC and Globecom, have papers in this area. Readers are encouraged to look into conference proceedings for the latest research results.

4.5 References

[1] F. Stajano and R. Anderson, "The resurrecting duckling: security issues for ad-hoc wireless networks," in *Proc. 7th International Workshop on Security Protocols, Lecture Notes in Computer Science*, Springer-Verlag, Berlin, Germany, Apr. 1999. Available from www.cl.cam.ac.uk/~fms27/duckling/duckling.html.

[2] F. Stajano, "The resurrecting duckling – what next?" *Proc. 8th International Workshop on Security Protocols, Lecture Notes in Computer Science*, Springer-Verlag, Berlin, Germany, April 2000. Available from www.cl.cam.ac.uk/~fms27/duckling/ duckling-what-next.html.

[3] N. Asokan and P. Ginzboorg, "Key agreement in ad hoc networks," *Computer Communications*, vol. 23, 2000, pp. 1627–1637.

[4] R. R. S. Verma, D. O'Mahony, and H. Tewari, *NTM – Progressive Trust Negotiation in Ad Hoc Networks*, www.cs.tcd.ie/Donal.Omahony/iei-ntm.pdf.

[5] S. Jacobs, *Mobile IP Public Key Based Authentication*, Internet draft, IETF, www3.tools.ietf.org/html/draft-jacobs-mobileip-pki-auth-02, Mar. 1999.

[6] A. Fasbender, D. Kesdogan, and O. Kubitz, "Variable and scalable security: protection of location information in mobile IP," *Mobile Technology for the Human Race, IEEE 46th Vehicular Technology Conference*, 1996.

[7] P. Eronen, C. Gehrmann, and P. Nikander, *Securing Ad Hoc Jini Services*, www.niksula.hut.fi/~peronen/publications/nordsec_2000.pdf, 2000.

[8] J. Kong, P. Zerfos, H. Luo, S. Lu, and L. Zhang, "Providing robust and ubiquitous security support for mobile ad-hoc networks," *9th IEEE International Conference on Network Protocols*, Riverside, CA, Nov. 2001, pp. 251–260.

[9] S. Mäki, M. Hietalahti, and T. Aura, *A Survey of Ad-Hoc Network Security*, Interim report of project 007 – Security of Mobile Agents and Ad-Hoc Societies, Helsinki University of Technology. Laboratory for Theoretical Computer Science, Sep. 2000.

[10] H. Luo and S. Lu, *Ubiquitous and Robust Authentication Services for Ad Hoc Wireless Networks*, Technical Report TR-200030, Dept. of Computer Science, UCLA, 2000.

[11] S. Jacobs and M. S. Corson, *MANET Authentication Architecture*, http:// ietfreport.isoc.org/all-ids/draft-jacobs-imep-auth-arch-00.txt, 1999.

[12] L. Venkatraman and D. P. Agrawal, "A novel authentication scheme for ad-hoc networks," *2nd IEEE Wireless Communications and Networking Conference*, Chicago, Sep. 2000.

[13] M. Jiang, J. Li, and Y. C. Tay, "*Cluster Based Routing Protocol*," http:// tools.ietf.org/html/draft-ietf-manet-cbrp-spec-01.txt, Aug 1999.

[14] S. Levijoki, *Authentication, Authorization and Accounting in Ad Hoc Networks*, www.tml.tkk.fi/Opinnot/Tik-110.551/2000/papers/authentication/aaa.htm, 2000.

[15] M. Blaze, J. Feigenbaum, and J. Lacy, "Decentralized trust management," *IEEE Symposium on Security and Privacy*, May 1996, pp. 164–173.

[16] M. Blaze, J. Feigenbaum, J. Ioannidis, and A. Keromytis, *The KeyNote Trust-Management System Version 2*, www.ietf.org/rfc/rfc2704.txt, Sep. 1999.

[17] J. Myers, *Simple Authentication and Security Layer (SASL)*, www.ietf.org/rfc/rfc2222.txt, Oct. 1997.

[18] D. Maughan, M. Schertler, M. Schneider, and J. Turner, *Internet Security Association and Key Management Protocol (ISAKMP)*, http://www.ietf.org/rfc/rfc2408.txt, Nov. 1998.

[19] D. Harkins and D. Carrel, *The Internet Key Exchange (IKE)*, www.ietf.org/rfc/rfc2409.txt, Nov. 1998.

5

Intrusion detection

Intrusion detection has, over the last few years, assumed paramount importance within the broad realm of network security; more so in the case of wireless ad hoc networks. These are networks that do not have an underlying infrastructure and the network topology is constantly changing. The inherently vulnerable characteristics of wireless ad hoc networks make them susceptible to attacks and countering attacks might end up being too little too late. Secondly, with so much advancement in hacking, if attackers try hard enough, they will eventually succeed in infiltrating the system. This makes it important to monitor constantly (or at least periodically) what is taking place on a system and look for suspicious behavior. Intrusion detection systems (IDSs) do just that: monitor audit data, look for intrusions to the system, and initiate a proper response (e.g., email the systems administrator, start an automatic retaliation, etc.). As such, there is a need to complement traditional security mechanisms with efficient intrusion detection and response. This chapter discusses the problem of intrusion detection in mobile ad hoc networks and presents the solutions that have been proposed so far.

5.1 Introduction

Wireless ad hoc networks have been in focus within the wireless research community. Essentially, these are networks that do not have an underlying fixed infrastructure. Mobile hosts "join" in, on the fly, and create a network on their own. With the network topology changing dynamically and the lack of a centralized network management functionality, these networks tend to be vulnerable to a number of attacks.

Mobile nodes within one another's radio range can communicate through wireless links and, thus, dynamically form a network. Wireless devices that are not in direct range communicate via intermediate devices, namely, multi-hop communication. Thus, an ad hoc network is a collection of autonomous nodes that form a dynamic, purpose-specific, multi-hop radio network in a decentralized fashion. The quintessential nature of such networks is the conspicuous absence of a fixed support infrastructure, such as mobile switching centers, base stations, access points, and other centralized machinery seen traditionally in wireless networks. The network topology is constantly changing as a result of nodes joining in and moving out. Packet forwarding, routing, and other network operations are carried out by the individual nodes themselves.

Wireless ad hoc networks find application in military operations in which planes, tanks, and moving personnel can communicate. Rescue missions and emergency services also find use in such networks. Other examples include virtual classrooms and conferences, wherein people can set up a network on the spot through their laptops, PDAs, and other mobile devices, assuming they share the same physical medium such as direct sequence spread spectrum (DSSS) or frequency hopped spread spectrum (FHSS).

The unreliability of wireless links between nodes, the constantly changing topology owing to the movement of nodes in and out of the network, and the lack of incorporation of security features in statically configured wireless routing protocols not meant for ad hoc environments all lead to increased vulnerability and exposure to attacks. Security in wireless ad hoc networks is particularly difficult to achieve, notably because of the limited physical protection to each of the nodes, the sporadic nature of connectivity, the absence of a certification authority, and the lack of a centralized monitoring or management unit. Intrusion prevention is not guaranteed to work all the time and this clearly underscores the need for intrusion detection as a frontline security research area under the umbrella of ad hoc network security. If an intrusion is detected quickly enough, the intruder can be identified and ejected from the system before any damage is done or any data are compromised. Moreover, an effective intrusion detection system can serve as a deterrent, so acting to prevent intrusions. Intrusion detection enables the collection of information about intrusion techniques that can be used to strengthen the intrusion prevention facility. In this chapter, I look at how ad hoc networks can be secured, to a certain extent, using traditional techniques. I then examine the different intrusion detection techniques proposed for these networks.

The rest of the chapter is organized as follows: in Section 5.2, I present the characteristics of wireless ad hoc networks, which make them so vulnerable to attacks. The fundamentals of intrusion detection are covered in Section 5.3, along with a classification of these systems. I then look at the requirements and characteristics of intrusion detection systems in Section 5.4. Section 5.5 – the piece de resistance – presents a state-of-the-art view of research in intrusion detection in the ad hoc environment. Section 5.6 is devoted to the comparison of different intrusion detection schemes against a set of attributes that are desirable in any intrusion detection scheme.

5.2 Security vulnerabilities in mobile ad hoc networks (MANETs)

There are various reasons why wireless ad hoc networks are at risk, from a security point of view. The next paragraphs discuss the characteristics that make these networks vulnerable to attacks.

In traditional wireless networks, mobile devices associate themselves with an access point, which is in turn connected to other wire-line machinery, such as a gateway or name server, which manage the network management functions. Ad hoc networks, on the other hand, do not have a centralized piece of machinery, such as a name server, which, if present as a single node, can be a single point of failure. The absence of infrastructure and, subsequently, the absence of authorization facilities impede the usual practice of establishing a line of defense, distinguishing nodes as trusted and non-trusted. There may be no grounds for an a priori classification, since all nodes are required to cooperate in supporting the network operation, while no prior security association (SA) can be assumed for all the network nodes. Freely roaming nodes form transient associations with their neighbors; they join and leave sub-domains independently with and without notice.

An additional problem related to the compromised nodes is that of the potential *Byzantine* failures encountered within mobile ad hoc network (MANET) routing protocols, wherein a set of the nodes could be compromised in such a way that the incorrect and malicious behavior cannot be directly noted at all. Such malicious nodes can also create new routing messages and advertize non-existent links, provide incorrect link state information and flood other nodes with routing traffic, thus inflicting Byzantine failures on the system.

The wireless links between nodes are highly susceptible to link attacks, which include passive eavesdropping, active interfering, leakage of secret information, data tampering, impersonation, message replay, message distortion, and denial of service. Eavesdropping might give an adversary access to secret information, violating confidentiality. Active attacks might allow the adversary to delete messages, to inject erroneous messages, to modify messages, and to impersonate a node, thus violating availability, integrity, authentication, and non-repudiation.

The presence of even a small number of adversarial nodes could result in repeatedly compromised routes, and, as a result, the network nodes would have to rely on cycles of timeout and new route discoveries to communicate. This would incur arbitrary delays before the establishment of a non-corrupted path, while successive broadcasts of route requests would impose excessive transmission overhead. In particular, intentionally falsified routing messages would result in a denial of service (DoS) experienced by the end nodes.

Moreover, the battery-powered operation of ad hoc networks gives attackers ample opportunity to launch a denial-of-service attack by creating additional transmissions or expensive computations to be carried out by a node in an attempt to exhaust its batteries.

Attacks against MANETs can be divided into two groups: passive attacks typically involve only eavesdropping of data, whereas active attacks involve actions performed by adversaries, for instance, the replication, modification, and deletion of exchanged data. External attacks are typically active attacks that are targeted to prevent services from working properly or to shut them down completely. Intrusion prevention measures, such as encryption and authentication, can only go so far as to prevent external nodes from disrupting the traffic, but can do little when compromised nodes internal to the network begin to disrupt traffic. Internal attacks are typically more severe, since malicious insider nodes already belong to the network as an authorized party and are thus protected by the security mechanisms that the network and its services offer. Thus, such compromised nodes, which may even operate in a group, may use the standard security means to actually protect their attacks.

In summary, a malicious node can disrupt the routing mechanism employed by several routing protocols. For example, it can:

(1) Attack the route discovery process by:
 • Changing the contents of a discovered route;
 • Modifying a route reply message, causing the packet to be dropped as an invalid packet;
 • Invalidating the route cache in other nodes by advertizing incorrect paths;
 • Refusing to participate in the route discovery process.
(2) Attack the routing mechanism by:
 • Modifying the contents of a data packet or the route via which that data packet is supposed to travel;
 • Behaving normally during the route discovery process but dropping data packets, causing a loss in throughput.
(3) Generate false route error messages whenever a packet is sent from a source to a destination.
(4) Launch denial of service attacks by:
 • Sending a large number of route requests. Because of the mobility aspect of MANETs, other nodes cannot determine whether the large number of route

requests are a consequence of a DoS attack or are due to a large number of broken links because of high mobility.

- Spoofing its IP and sending route requests with a fake ID to the same destination, causing a DoS at that destination.

The above discussion makes it clear that ad hoc networks are inherently insecure, more so than their wire-line counterparts, and need intrusion detection schemes before it is too late to counter an attack. If there are attacks on a system, one would like to detect them as soon as possible (ideally in real time) and take appropriate action. This is essentially what an IDS does. I now discuss the basics of an intrusion detection system and provide a classification of such systems.

5.3 Intrusion detection systems: a brief overview

Intrusion detection can be defined as the automated detection and subsequent generation of an alarm to alert the security apparatus at a location if intrusions have taken place or are taking place. An intrusion detection system is a defense system, which detects hostile activities in a network and then tries to prevent such activities that may compromise system security. Intrusion detection systems achieve detection by continuously monitoring the network for unusual activity. The prevention part may involve issuing alerts as well as taking direct preventive measures, such as blocking a suspected connection. In other words, intrusion detection is a process of identifying and responding to malicious activity targeted at computing and networking resources. In addition, IDS tools are capable of distinguishing between insider attacks originating from inside the network and external ones. Unlike firewalls, which are the first line of defense, intrusion detection systems come into the picture only after an intrusion has occurred and a node or a network has been compromised. That is why intrusion detection systems are aptly called the second line of defense.

Generally speaking, the following security related features are not part of the intrusion detection system. An IDS:

(1) is NOT an anti-virus system, designed to detect malicious software, such as viruses, Trojans, worms, etc.
(2) is NOT a network logging system used, for example, to detect complete vulnerability to any denial-of-service (DoS) attack across a congested network. These are network traffic monitoring systems.
(3) is NOT a vulnerability assessment tool that checks for bugs and flaws in operating systems and network services. Such an activity would fall under the purview of security scanners.

A basic model of an intrusion detection system is likely to include quite a few elements. Primarily, intrusion detection decisions are based on the collected audit data. Sources of data can include keyboard input, command-based logs and application-based logs. Audit data are stored either indefinitely, for later reference, or temporarily, to await processing. The enormous volume of data makes this a crucial element in intrusion detection systems. One or many algorithms are executed to find evidence in the audit trail of suspicious behavior. An IDS is generally controlled by the configuration settings that would specify how and where to collect audit data, how to respond to intrusions, etc. Access to these configuration settings would give a potential intruder vital information on which avenues of attack are likely to go undetected. Reference data store information about known intrusion signatures (for misuse systems) or profiles of normal behavior (for anomaly systems). The processing element must frequently store intermediate results, an example of which might be information about partially fulfilled intrusion signatures. The space needed to store this active data can grow quite large too. And, finally, the alarm part of the system handles all output from the system. Examples include automated response to suspicious activity and notification to the user.

Intrusion detection can be classified into three broad categories: anomaly detection, signature or misuse detection, and compound detection. I discuss each of these as per the taxonomy proposed in [1]:

(1) *Anomaly detection*: In an anomaly detection system, a baseline profile of normal system activity is created. Any system activity that is a deviation from the baseline is treated as a possible intrusion. The problems with strict anomaly detection are that (1) anomalous activities that are not intrusive are flagged as intrusive and (2) intrusive activities that are not anomalous result in false negatives. One disadvantage of anomaly detection for mobile computing is that the normal profile must be periodically updated and deviations from the normal profile must be computed. The periodic calculations can impose a heavy load on some resource-constrained mobile devices and perhaps a lightweight approach that involves comparatively less computation might be better suited.

(2) *Misuse detection*: In misuse detection, decisions are made on the basis of knowledge of a model of the intrusive process and what traces it ought to leave in the observed system. *Legal* or *illegal* behavior can be defined and observed behavior can be compared accordingly. Such a system tries to detect evidence of intrusive activity irrespective of any knowledge regarding the background traffic, i.e., the normal behavior of the system.

(3) *Specification-based detection*: Specification-based detection defines a set of constraints that describe the correct operation of a program or protocol, and monitors the execution of the program with respect to the defined constraints. This

technique may provide the capability to detect previously unknown attacks, while exhibiting a low false positive rate.

An offshoot of misuse and anomaly detection is compound detection, which is a misuse inspired system that forms a compound decision in view of a model of both the normal behavior of the system and the intrusive behavior of the intruder. The detector operates by detecting the intrusion against the background of the normal traffic in the system. These detectors have a much better chance of correctly detecting truly interesting events in the supervized system, since they both know the patterns of intrusive behavior and can relate them to the normal behavior of the system. They would, at the very least, be able to qualify their decisions better.

5.3.1 Intrusion response

The type of intrusion response for wireless ad hoc networks depends on the type of intrusion, the network protocols and applications in use, and the confidence (or certainty) in the evidence. A few likely responses include:

(1) Reinitializing communication channels between nodes (e.g., force re-key);
(2) Identifying the compromised nodes and reorganizing the network to preclude the compromised nodes;
(3) The IDS agent informing the end user, who may, in turn, conduct an independent investigation and take appropriate action;
(4) Initiate a "re-authentication" request to all nodes in the network to prompt the end users to authenticate themselves (and hence their wireless nodes), using out-of-band mechanisms (like visual contacts). Only the re-authenticated nodes, which may collectively negotiate a new communication channel, will recognize each other as legitimate. That is, the compromised or malicious nodes can be excluded.

5.4 Requirements for an intrusion detection system for mobile ad hoc networks

There are two key requirements that any IDS needs to fulfill. These are *effectiveness* – how to make the intrusion detection system classify malign and benign activity correctly – and *efficiency* – how to run the intrusion detection system in a cost-effective manner as far as possible. In other words, these two requirements in essence suggest that an IDS should detect a substantial percentage of intrusions into the supervized system, while still keeping the false alarm rate at an acceptable level at a lower cost. It is expected that an ideal IDS is likely to support several of the following requirements:

(1) The intrusion detection system should not introduce a new weakness in the MANET. That is, the IDS itself should not make a node any weaker than it already is.
(2) An intrusion detection system should run continuously and remain transparent to the system and the users.
(3) The intrusion detection system should use as little of the system resources as possible to detect and prevent intrusions. Intrusion detection systems that require excessive communication among nodes or run complex algorithms are not desirable.
(4) It must be fault tolerant in the sense that it must be able to recover from system crashes, hopefully recover to the previous state, and resume the operations before the crash.
(5) Apart from detecting and responding to intrusions, IDS should also resist subversion. It should monitor itself and detect whether it has been compromised by an attacker.
(6) An intrusion detection system should have a proper response. In other words, an IDS should not only detect but should also respond to the detected intrusions, preferably without human intervention.
(7) Accuracy of the intrusion detection system is another major factor in MANETs. Fewer false positives and false negatives are desired.
(8) It should inter-operate with other intrusion detection systems collaboratively to detect intrusions. For example, the IETF Intrusion Detection Working Group (IDWG) [2] is working towards proposing such a specification.

5.5 Intrusion detection in MANETs

A great deal of research work has already been carried out in intrusion detection for traditional wired networks. However, applying the research of wired networks to wireless networks is not an easy plug-and-play task because of key architectural differences, principal among them being the lack of fixed infrastructure. The absence of a physical infrastructure facilitates the attacker's task, since it is easier to eavesdrop on network traffic in a wireless environment.

Wireless ad hoc networks, due to their vulnerabilities, provide a tougher challenge for designing an IDS. Without centralized audit points, such as routers, gateways, etc., an IDS for ad hoc networks is limited to use only the current traffic coming in and out of the node as audit data. Another key requirement is that the algorithms that the IDS uses must be distributed in nature, and should take into account the fact that a node can only see a portion of the network traffic. Moreover, since ad hoc networks are dynamic and nodes can move about freely, there is a possibility that one or more nodes could be captured and compromised, especially if the network is in a hostile environment.

If the algorithms of the IDS are cooperative, it becomes important to be skeptical of which nodes one can trust. Therefore, intrusion detection systems on ad hoc networks have to be wary of attacks made from nodes in the network itself, not just attacks from outside the network. Also, mobile networks cannot communicate as frequently as their wired counterparts to detect intrusions, as they must conserve bandwidth resources. Bandwidth and other issues, such as battery life, compound the problem even further. The availability of partial audit data makes it harder to distinguish an attack from regular network use.

In this section, a state-of-the-art view of research in intrusion detection systems for MANETs including proposed architectures and development work that is in progress are presented.

5.5.1 A distributed IDS

In their pioneering work on intrusion detection in MANETs, the authors Zhang and Lee describe a distributed and cooperative intrusion detection model, where every node in the network participates in intrusion detection and response [3]. In this model, an IDS agent runs at each mobile node and performs local data collection and local detection, whereas cooperative detection and global intrusion response can be triggered when a node reports an anomaly. The authors consider two attack scenarios separately: (i) abnormal updates to routing tables, and (ii) detecting abnormal activities in layers other than the routing layer.

An IDS agent is structured into six pieces, as shown in Fig. 5.1. Each node detects local intrusion independently and neighboring nodes work collaboratively on a larger scale. Individual IDS agents placed on each and every node run independently to monitor local activities (including user, systems, and communication activities within the radio range), detect intrusions from

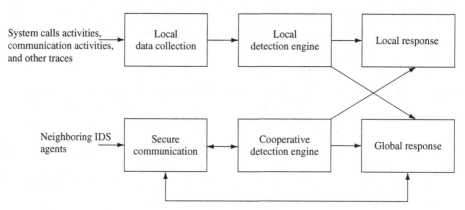

Figure 5.1 Intrusion detection system for MANETs

local traces, and initiate responses. Neighboring IDS agents cooperatively participate in global intrusion detection actions when an anomaly is detected in local data or if there is inconclusive evidence. The data collection module gathers local audit traces and activity logs, which are used by the local detection engine to detect local anomalies. Detection methods that need broader data sets or that require collaboration among local IDS agents use the cooperative detection engine. Both the local response and global response modules provide intrusion response actions. The local response module triggers actions local to this mobile node, for example, an IDS agent alerting the local user, while the global response module coordinates actions among neighboring nodes, such as the IDS agents in the network, electing a remedial action. A secure communication module provides a high-confidence communication channel among IDS agents.

The main contribution of [3] is that it presents a distributed and cooperative intrusion detection architecture, which is based on statistical anomaly detection techniques. This paper was among the first that had such a detailed distributed design. The design of actual detection techniques, their performance as well as verification, however, have not been addressed in the paper.

5.5.2 AODV protocol-based IDS

Bhargava and Agrawal [4] proposed an intrusion detection and response model (IDRM) to enhance security in the ad hoc on demand distance vector (AODV) routing protocol [5]. The intrusion detection model proposed by the authors is an extension of the model described in Section 5.1.

Figure 5.2 illustrates how the IDRM provides security to AODV protocol. In this scheme, each node employs an IDRM that utilizes neighborhood information to detect misbehavior of its neighbors. When the misbehavior count for a node exceeds a predefined threshold, the information is sent out to other nodes as part of global response. The other nodes receive this information, check their local "malcount" for this malicious node, and add their results to the initiator's response. In the intrusion response model (IRM), a node identifies that another node has been compromised when its malcount increases

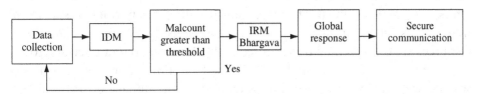

Figure 5.2 Handling of internal attacks

beyond the threshold value for that allegedly compromised node. In such cases, it propagates this information to the entire network by transmitting a special type of packet called a "MAL" packet. If another node also suspects that the node that has been detected as compromised, it reports its suspicion to the network and retransmits another special type of packet, called "REMAL." If two or more nodes report suspicions about a particular node, another of the special packets, called a "PURGE" packet, is transmitted to isolate the malicious node from the network. All nodes that have a route through the compromised node look for newer routes. All packets received from a compromised node are dropped.

Examples of the internal attacks are distributed false route request, DoS, impersonation, and compromise of a destination. The authors have proposed the following ways of identifying these internal attacks:

5.5.3 Techniques for intrusion-resistant ad hoc routing algorithms (TIARA)

Techniques for intrusion-resistant ad hoc routing algorithms (TIARA) are a set of innovative design techniques that strengthen MANETs against denial of service attacks, produced at the Architecture Technology Corporation [6]. The TIARA mechanisms limit the damage sustained by MANETs from intrusion attacks and allow for continued network operation at an acceptable level during such attacks. They provide protection against attacks on control routing traffic as well as data traffic, thereby providing a comprehensive defense against intruders. Because of routing algorithm independence, they allow for widespread applicability and support secure enclaves for dynamic coalitions.

The TIARA approach uses fully distributed lightweight firewalls for ad hoc wireless networks, distributed traffic policing mechanisms, intrusion tolerant routing, distributed intrusion detection mechanisms, flow monitoring, reconfiguration mechanisms, multi-path routing, and source-initiated route switching. The flow-based route access control (FRAC) rules define admissible flows. Per-flow security association is instantiated by a secure session set-up signaling protocol and contains information for packet authentication. Also, fast authentication enables low-overhead integrity checks on packet flow-IDs and sequence numbers. There is referral-based resource allocation, which limits a network's exposure to resource usurpation by spurious sessions, and flows are assigned an initial allowable resource usage. Moreover, additional resources are only granted if the source of the flow can present referrals from a certain number of trusted nodes. Referrals have time-bound validity. Flow-specific sequence numbers to limit and contain the impact of traffic replay attacks are embedded in secret locations within each packet. The destinations

Table 5.1 *A summary of TIARA countermeasures against intrusion attacks*

Intrusion attacks → Countermeasures ↓	Spurious traffic	Packet replay	Session flooding	Flow disruption	Route hijacking
FRAC	×				
Fast authentication	×	×	×		
Sequence numbers		×			
Referrals			×		
Flow monitoring				×	×
Multi-path routing				×	×
Source initiation route				×	×
Switching					

of flow monitors select flow parameters to detect intrusion-induced path failures, and multipath-routing and source-initiated route switching diverts flow through available alternate paths to circumvent an intruder. Efforts are being made to implement dynamic, on-the-fly modifications to FRAC (firewall) policies; real-time-referral-based resource allocation; lightweight implementation of traffic policing; fast authentication mechanisms that are resistant to traffic analysis; embedding sequence numbers; and path labels in encrypted packets. Though the proposed architecture seems to cover most of the important aspects of intrusion detection and prevention in MANETs, implementation of such a design methodology entails the extensive modification of the routing algorithms in a MANET. A summary of countermeasures used in TIARA against intrusion attacks in shown in Table 5.1.

5.5.4 Watchdog-pathrater approach

Sergio Marti *et al.* discussed two techniques that improve throughput in MANETs in the presence of compromised nodes that agree to forward packets but fail to do so [7]. A node may misbehave because it is overloaded, selfish, malicious, or broken. An overloaded node lacks the CPU cycles, buffer space, or available network bandwidth to forward packets. A selfish node is unwilling to spend battery life, CPU cycles, or available network bandwidth to forward packets not of direct interest to it, even though it expects others to forward packets on its behalf. A malicious node launches a denial of service attack by dropping packets. A broken node might have a software fault that prevents it from forwarding packets.

To mitigate the decrease in the throughput due to the above node categories, the authors use watchdogs that identify misbehaving nodes and a pathrater that helps routing protocols to avoid these nodes. When a node forwards a packet,

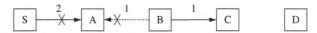

Figure 5.3 Node A does not hear node B forward packet 1 to node C, because node B's transmission collides at node A with packet 2 from source S

Figure 5.4 Node A believes that node B has forwarded packet 1 to C, though node C never received the packet due to collision with packet 2

the node's watchdog verifies that the next node in the path also forwards the packet. The watchdog does this by listening promiscuously to the next node's transmissions. If the next node does not forward the packet, then it is misbehaving. Every time a node fails to forward the packet, the watchdog increments a failure tally. If the tally exceeds a certain threshold, the watchdog determines that the node is misbehaving and this node is then avoided using the pathrater. The pathrater, run by each node in the network, combines knowledge of misbehaving nodes with link reliability data to pick the route most likely to be reliable. Each node maintains a rating for every other node it knows about in the network. It calculates a path metric by averaging the node ratings in the path.

The watchdog technique has its own advantages and weaknesses. The dynamic source routing (DSR) [8] with the watchdog has the advantage that it can detect misbehavior at the forwarding level and not just the link level. The watchdog's weaknesses are that it might not detect a misbehaving node in the presence of:

(1) Ambiguous collisions: prevents node A from overhearing the transmission from node B, as shown in Fig. 5.3;
(2) Receiver collisions: node A can only tell whether B has sent a packet, but it cannot tell if node C received it or not, as shown in Fig. 5.4.
(3) Limited transmission power: a misbehaving node could limit its transmission power such that the signal is strong enough to be overheard by the previous node but too weak to be received by the true recipient;
(4) False misbehavior: this occurs when a node falsely reports other nodes as misbehaving;
(5) Partial dropping: a node can circumvent the watchdog by dropping packets at a lower rate than the watchdog's configured minimum misbehaving threshold.

5.5.5 Anomaly detection for mobile wireless networks

An anomaly detection architecture that was proposed in [9] is shown in Fig. 5.5. In this scheme every node in the mobile ad hoc network participates in intrusion detection and response. Every node is responsible for detecting

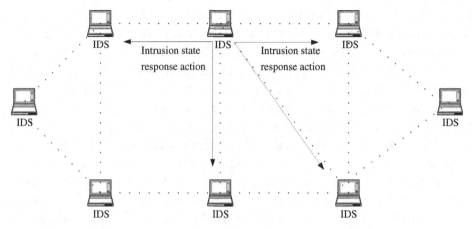

Figure 5.5 Intrusion detection system (IDS) architecture for wireless ad hoc networks

signs of intrusion locally and independently by monitoring activities such as user and system activities and the communication activities within the radio range, but neighboring nodes can investigate a broader range collaboratively. The internal structure of the detection scheme is shown conceptually in Fig. 5.1. Information-theoretic measures [10], such as entropy and conditional entropy, are used to describe the characteristics of normal information flows and classification algorithms are used to build anomaly detection models. For example, a classifier trained using normal data can be used to predict the next event, given the previous n events. In monitoring, when the actual event is not what the classifier has predicted, there is an anomaly. When constructing a classifier, features with high information gain (or reduction in entropy) are needed. That is, a classifier needs feature value tests to partition the original (mixed and high entropy) dataset into pure (and low entropy) subsets, each ideally with one (correct) class of data.

Using the above mentioned framework, the following procedure is utilized for anomaly detection:

(1) Select (or partition) audit data so that the normal dataset has low (conditional) entropy;
(2) Perform appropriate data transformation according to the entropy measures (e.g., by constructing new features with high information gain);
(3) Compute classifier using training data;
(4) Apply the classifier to test data; and
(5) Post-process alarms to produce intrusion reports.

Local routing information, including cache entries and traffic statistics, are used as an audit data source because remote nodes can be compromised and their data cannot be trusted. Since classifiers are used as detectors there is a need to select or construct features from the available audit data that have high information gain. An unsupervized method is used to construct the feature set. First, a large feature set is constructed to cover a wide range of behaviors. Then a small number of training runs can be performed with the whole set of features on small audit data traces randomly chosen from previously stored audit logs. For each training run, a corresponding model is built. The features that appear in the models and have weights not smaller than a minimum threshold are selected into the essential feature set. For different routing protocols and different scenarios, the essential feature set is different. In practice, the feature set needs to be updated after a certain period, as the characteristics of routing behavior can change with time. The heuristic is that with sufficiently high dimension, data can be separated by a hyper-plane, thus achieving the classification goal. Given an execution trace, a detector is first applied to examine each observation. Then a post-processing scheme is used to examine the predictions and generate intrusion reports. A detection model can make spurious errors and these false alarms should be filtered out. In contrast, a true intrusion session has "locality," i.e., it tends to result in many alarms within a short time window. Therefore, these alarms can be grouped into a single intrusion report.

5.6 Mobile agents for intrusion detection and response in MANETs

Mobile agents are a special kind of agent that have the ability to move through large networks. In moving, the agents can interact with nodes, collect information, and execute tasks assigned to them. Mobile agents offer several advantages such as reduction in the network load as well as latency, which is achieved by eliminating the need to move large amounts of data through the network by moving the analysis programs closer to the audit data. When portions of an intrusion detection system get destroyed or separated due to the network partitioning, the mobile agents can still continue to work, thereby increasing the fault tolerance level of the network. The mobile agents tend to be independent of platform architectures, therefore rendering agent-based intruder detection systems to run under different operating system environments.

5.6.1 Local intrusion detection system (LIDS)

The local intrusion detection system (LIDS) is distributed in nature and utilizes mobile agents on each of the nodes of the ad hoc network [11]. To make local intrusions a global concern for the entire network, the systems

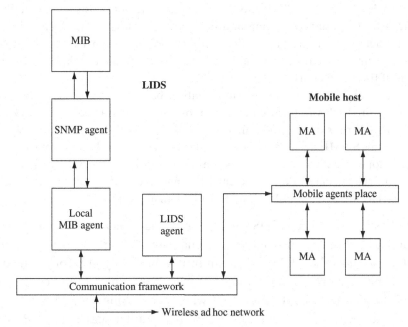

Figure 5.6 The LIDS architecture

existing on different nodes collaborate. The collaboration among the nodes is achieved using two types of data: security data to obtain complementary information from collaborating hosts and intrusion alerts to inform others of a locally detected intrusion. The LIDS has chosen to use SNMP (simple network management protocol) data located in MIBs (management information bases) as the audit source because SNMP offers several advantages, principal among them being that the cost of local information collection is negligible, if an SNMP agent is running on a node. Mobile agents (which need to be autonomous and adaptive) are used to transport SNMP requests to remote hosts to overcome the unreliability of SNMP message transfer over the UDP. A LIDS can delegate a specific mission to an agent, which will carry out its assigned task in an autonomous and asynchronous manner without any help from its LIDS.

The LIDS architecture is shown in Fig. 5.6. The key elements of the architecture are:

(1) A common communication framework to facilitate all external and internal communication with a LIDS;
(2) Several data collecting agents for different tasks, e.g.:
   • A local LIDS agent that is in charge of local intrusion detection and response and also responsible for reacting to intrusion alerts provided by other nodes in order to protect itself against this intrusion.

- Mobile agents that collect and process data on remote hosts with an ability to transfer the results of a computation back to their home LIDS or to migrate to another node for further investigation. The mobile agent place is responsible for the security control of these agents, but an agent should also be able to protect itself from malicious mobile agent places.
- The local MIB (management information base) agent provides a means of collecting MIB variables, either for mobile agents or for the local LIDS agent. If SNMP runs on the node the local MIB agent will be the interface with the running SNMP agent. For other scenarios, an SNMP based agent has to be developed to allow optimized updates and retrieval of the MIB variables used by intrusion detection. The local MIB agent would in that case act as an interface between the LIDS and this tailor-made agent.

In this design the local LIDS agent could use either misuse or anomaly detection as an intrusion detection mechanism. As far as response is concerned, as soon as the LIDS detects an intrusion locally, it informs the other nodes of the network. Locally, the node is empowered to refuse connections with the suspicious node, to exclude it when performing cooperative actions, or to exclude it from its community until it re-authenticates itself. By being informed of intrusions on remote hosts, the LIDS can act as a security tool and prevent the intruder from attacking it. The system designers recommend that for the best security in an ad hoc network, all the systems on nodes should run and cooperate continuously.

5.6.2 Intrusion detection architecture based on a static stationary database

A distributed IDS has been proposed at Mississippi State University, in which each node on the network has an IDS agent running on it [12]. The IDS agents on each node in the network work together via a cooperative intrusion detection algorithm to decide when and how the network is being attacked. The architecture is divided into parts: the mobile IDS agent, which resides on each node in the network, and the stationary secure database, which contains global signatures of known misuse attacks and stores patterns of each user's normal activity in a non-hostile environment.

Mobile IDS agents Each node in the network will have an IDS agent running on it all the time. This agent is responsible for detecting intrusions based on local audit data and participates in cooperative algorithms with other IDS agents to decide if the network is being attacked. Each agent has five parts: the local audit trail, the local intrusion database (LID), the secure communication module, the anomaly detection modules (ADMs), and the misuse detection modules (MDMs), as shown in Fig. 5.7.

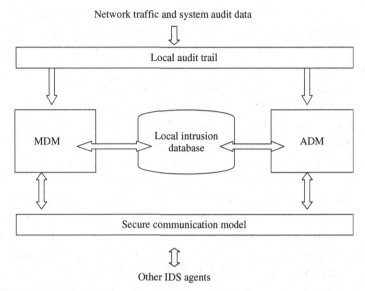

Figure 5.7 Proposed IDS based on a stationary secured database

- The *local intrusion database* (LID) is a local database that warehouses all information necessary for the IDS agent, such as the signature files of known attacks, the established patterns of users on the network, and the normal traffic flow of the network. The anomaly detection modules and misuse detection modules communicate directly with the LID to determine whether an intrusion is taking place.
- The *secure communication module* is necessary to enable an IDS agent to communicate with other IDS agents on other nodes. It will allow the MDMs and ADMs to use cooperative algorithms to detect intrusions. It may also be used to initiate a global response when an IDS agent or a group of IDS agents detects an intrusion. Data communicated via the secure communication module need to be encrypted.
- The *anomaly detection modules* (ADMs) are responsible for detecting a different type of anomaly. There can be any number of anomaly detection modules on each mobile IDS agent, each working separately or cooperatively with other ADMs.
- The *misuse detection modules* (MDMs) identify known patterns of attacks that are specified in the local intrusion database. Like the ADMs, if the audit data available locally are sufficient to determine whether an intrusion is taking place, the proper response can be initiated. It is also possible for a misuse detection module to use a cooperative algorithm to identify an intrusion.

Stationary secure database The stationary secure database (SSD) acts as a secure, trusted repository for mobile nodes to obtain information about the latest misuse signatures and to find the latest patterns of normal user activity. It is assumed that the attacker will not compromise the stationary secure database, as it is stored in an area of high physical security. The mobile IDS agents will collect and store audit data (such as user commands, network traffic, etc.) while in the field, and will transfer this information when they are attached to the SSD. The SSD will then use this information for data mining of new anomaly association rules. The SSD will also be the place where the system administrator can specify the newest misuse signatures. When the IDS agents are connected to the SSD, they will gain access to the latest attack signatures automatically. Using the SSD to communicate the new attack signatures and to establish new patterns of normalcy limits the amount of communication that must take place between IDS agents in the mobile ad hoc network. Despite all the benefits of having a stationary secure database in a mobile IDS architecture, there are a few disadvantages of relying on a stationary database to provide vital IDS information. If a stationary secure database is used, mobile nodes will have to be attached to the non-mobile database periodically to stay up-to-date with the latest intrusion information. This may not be an option for some mobile, ad hoc environments. Also, since the SSD must be a trusted source, it cannot be taken on site without significant risk.

5.6.3 Distributed intrusion detection using mobile agents

Kachirski and Guha have proposed a distributed intrusion detection system for ad hoc wireless networks based on mobile agent technology [13]. By efficiently merging audit data from multiple network sensors, their bandwidth-conscious scheme analyzes the entire ad hoc wireless network for intrusions at multiple levels, tries to inhibit intrusion attempts and provides a lightweight, low-overhead mechanism based on a mobile agent concept. There is an efficient distribution of mobile agents with specific IDS tasks according to their functionality across a wireless ad hoc network. The agents used are dynamically updateable, have limited functionality and can be viewed as components of a flexible, dynamically configurable intrusion detection system. Additionally, this scheme restricts computation-intensive analysis of the overall network security state to a few key nodes. These nodes are dynamically elected, and overall network security is not entirely dependent on any particular node. The modular approach taken has such advantages as increased fault tolerance, communications cost reduction, improved performance of the entire network, and scalability.

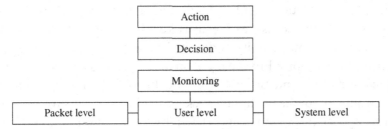

Figure 5.8 Modular intrusion detection architecture

The proposed intrusion detection system is built on a mobile agent framework, as shown in Fig. 5.8. It is a non-monolithic system and employs several sensor types that perform specific functions, such as:

- *Network monitoring* Only certain nodes have sensor agents for network packet monitoring, to preserve total computational power and battery power of mobile hosts.
- *Host monitoring* Every node on the mobile ad hoc network is monitored internally by a host-monitoring agent. This includes monitoring system-level and application-level activities.
- *Decision-making* Every node decides on its intrusion threat level on a host-level basis. Certain nodes collect intrusion information and make collective decisions about the network-level intrusions.
- *Action* Every node has an action module responsible for resolving intrusion situations on a host.

There are three major agent categories – monitoring, decision making, and action agents. Some are present on all mobile hosts, while others are distributed to only a selected group of nodes. While all the nodes accommodate host-based monitoring sensors of the IDS, a distributed algorithm is utilized to assign a few nodes to host sensors that monitor network packets and agents that make decisions. The mobile network is logically divided into clusters with a single cluster head for each cluster that monitors packets within the cluster. The selected nodes host network-monitoring sensors that collect all packets within the communication range, and analyze the packets for known patterns of attacks. Monitoring agents are categorized into packet monitoring sensors, user-activity sensors and system-level sensors. Local detection agents are located on each node of an ad hoc network, and act as user-level and system-level anomaly-based monitoring sensors. These agents look for suspicious activities on the host node, such as unusual process memory allocations, CPU activity, I/O activity, user operations (invalid login attempts with a certain pattern, super-user actions, etc.). If an anomaly is detected with strong evidence, a local detection agent will terminate the suspicious process or lock out the user and initiate re-issue of security keys for the entire network. If some inconclusive anomalous activity is detected on a host node by a monitoring agent, the node is reported to the decision agent of the same cluster of which the suspicious node is a member. If more conclusive evidence is gathered about this node from any source (including packet monitoring results from a network-monitoring agent), the action is undertaken by the agent on that node.

Decision agents are located on the same nodes as packet-monitoring agents. A decision agent contains a state machine for all the nodes within the cluster it resides in. As intrusion or anomalous activity evidence is gathered for each node, the agent can decide that a node has been compromised by looking at reports from the node's own local monitoring agents, and the packet-monitoring information pertaining to that node. When a certain level of threat is reached for a node in question, the decision agent dispatches a command that an action must be undertaken by the local agents on that node. In time, the threat level decreases for each node in the decision agent's database.

5.7 Summary

In this chapter, several intrusion detection schemes that have been proposed recently were surveyed. The main features of these schemes are summarized in Table 5.2. Severe memory constraints on a mobile device imply that misuse detection systems that need to store attack signatures will be relatively difficult to build and are likely to be less effective. Distributed anomaly detection, therefore, is by far the methodology of choice for intrusion detection in MANETs, and Table 5.2 clearly makes that point.

In Table 5.3, I compare different intrusion detection systems presented in this paper with the attributes of an ideal IDS. These attributes are: fault tolerance, scalability, inter-operability with other intrusion detection systems, ability to detect new attack patterns, and whether the proposed system introduces new weaknesses in terms of excessive overheads in terms of communication, storage, energy, or computation overheads.

We see from Table 5.3 that there is a trend to use mobile agents for intrusion detection and response in mobile ad hoc networks because these agents address the search and analysis problems involving multiple distributed resources in an efficient manner. As indicated by column 4, most proposed systems lack inter-operability because they do not use the common message format, e.g. one that has been proposed by the IETF for communication between various IDS agents. Inter-agent communication security is another area in which many of the proposed systems do not fare well.

Table 5.2 *Summary of proposed intruder detection systems*

Proposed system	Main features	Methodology
Distributed intrusion detection system for ad hoc networks	Statistical anomaly detection to detect local and global intrusions.	Distributed anomaly detection
Intrusion detection and response model for the AODV protocol	Collaborative-threshold-based scheme, where nodes watch neighbors for malicious activity. If two or more nodes report about a particular node, the malicious node is isolated from the network.	Distributed anomaly detection
Techniques for intrusion resistant ad hoc routing algorithms (TIARA)	Presents general design principles and techniques independent of routing algorithm that can be incorporated in MANETs for robust fault tolerant networks.	Distributed anomaly detection
Watchdog and pathrater	A threshold based scheme where nodes watch neighbors for signs of malicious activity. Once a threshold is crossed, the malicious nodes are excluded from the network.	Distributed anomaly detection
Local intrusion detection system	Mobile agents use local SNMP data located in the management information base as audit sources for intrusion detection. Also implemented is the use of the intrusion detection message exchange format (IDMEF) and a protocol for transporting such alerts (intrusion detection exchange protocol (IDXP)) to ensure that intrusion detection systems running on a broad range of platforms can still interact and exchange intrusion related information.	Mobile-agents-based distributed anomaly detection
Intrusion detection architecture based on a secure stationary database	Secure, centralized, and stationary database used to store misuse signatures and user profiles. Uses mobile agents for intrusion detection.	Mobile-agents-based compound detection
Distributed intrusion detection system based on mobile Agent	Audit data from multiple sensors used to implement a bandwidth conscious scheme for distributed intrusion detection using mobile agents.	Mobile-agents-based anomaly detection

Table 5.3 *Comparison of different proposed architectures against ideal characteristics for IDS in MANETs*

Proposed system	Attributes of an ideal intrusion detection system for MANETs				
	Introduces weaknesses and overheads?	Fault tolerant?	Inter-operates with other IDSs?	Scalable?	Detects new attack patterns?
Distributed intrusion detection system for ad hoc networks	Yes. The use of encryption for communication between IDS agents will slow down the communication process.	Yes	No	Yes	Yes, but is ineffective against IP spoofing.
Intrusion detection and response model for the AODV protocol	Yes. The use of encryption for communication between IDS agents will slow down the communication process.	Yes	No	Yes	Yes
Techniques for intrusion resistant ad hoc routing algorithms (TIARA)	No	Yes	No	Yes	Yes
Watchdog and pathrater	No	Yes	No	Yes	Yes, but fails against collaborative attacks.
Local intrusion detection system	No	No. Security of the agent platform itself is not addressed.	Yes	Yes	Yes
Intrusion detection architecture based on a secure stationary database	Yes. Limits the communication overhead but might increase the storage overhead in terms of misuse signatures on a node	No. the stationary static database could be a bottleneck or single point of failure.	No	Yes	Yes, but with a delay. The stationary database has to mine the limited audit data given to it by agents and has to broadcast new signatures.
Distributed intrusion detection system based on mobile agent	No	Yes	No	Yes	Yes

Taking a leaf out of Axelsson's survey on intrusion detection systems [1] for wire-line networks, we see that the following dichotomies in system characteristics hold true for wireless ad hoc networks.

- *Time of detection* Two main groups can be identified in wireless ad hoc networks too: those that attempt to detect intrusions in real time and those that process audit data with some delay.
- *Locus of data processing* The audit data in general are processed and new rules are derived from them in a distributed fashion. Each node in most of the surveyed systems takes the distributed approach to avoid being a single point of failure. The intrusion detection architecture based on a secure stationary database is the only exception, where audit data are transferred to the stationary secure database with the help of mobile agents, and these audit data are then mined for new misuse patterns.
- *Security* The ability to withstand a hostile attack against the intrusion detection system itself. This area has been the subject of little investigation. With the trend towards using mobile agents for intrusion detection, most of the surveyed systems that use mobile agents still do not consider the security of the agent platform itself.
- *Degree of inter-operability* The degree to which the system can inter-operate in conjunction with other intrusion detection systems and accept audit data and reports from different sources. This is not the same as the number of different platforms on which the intrusion detection system itself runs. With the exception of one, most of the proposed systems are not inter-operable with each other.

5.8 Further reading

This chapter presents the current state of the art in the area of intrusion detection and response for wireless mobile ad hoc networks. Even though research in intrusion detection started at least fifteen years ago in the wired world, its application to wireless ad hoc networks is a rather recent development. Wireless ad hoc networks are intrinsically resource constrained, and this makes several of the schemes proposed in the wired world inadequate, as discussed earlier. Approaches that require analysis of large trace data or attack signatures (used by misuse detection techniques) or require centralized analysis engines are not preferable. Instead the schemes that are distributed and collaborative, e.g., anomaly-detection-based schemes, are likely to be more applicable. One key advantage of using an anomaly detection scheme is that it requires less modification of current routing protocols and allows the trace analysis and anomaly detection to be performed locally in each node. At present, IETF has a working group on intrusion detection [2] that covers the current and future research topics and is an excellent source for the information on the scope of future work. Readers should refer to [14, 15, and 16] for additional reading material on intrusion detection for wireless ad hoc networks.

5.9 References

[1] S. Axelsson, *Intrusion Detection Systems: A Taxomomy and Survey*, Technical report no. 99–15, Dept. Computer Engineering, Chalmers University of Technology, Sweden, Mar. 2000.

[2] *Intrusion Detection Exchange Format (idwg)*, www.ietf.org/html.charters/OLD/idwg-charter.html, 2005.

[3] Y. Zhang and W. Lee, "Intrusion detection in wireless ad-hoc networks," *6th International Conference on Mobile Computing and Networking (MOBICOM'00)*, Aug. 2000, pp. 275–283.

[4] S. Bhargava and D. P. Agrawal, "Security enhancements in AODV protocol for wireless ad hoc networks," *54th Vehicular Technology Conference*, vol. 4, 7–11 Oct. 2001, pp. 2143–2147.

[5] C. E. Perkins, E. M. Royer, and S. R. Das, *Ad Hoc On-demand Distance Vector Routing*, Oct. 1999 IETF Draft.

[6] R. Ramanujan, A. Ahamad, J. Bonney, R. Hagelstrom, and K. Thurber, "Techniques for intrusion-resistant ad hoc routing algorithms (TIARA)," *21st Century Military Communications Conference Proceedings*, vol. 2, 22–25 Oct. 2000, pp. 660–664.

[7] S. Marti, T. J. Giuli, K. Lai, and M. Baker, "Mitigating routing misbehavior in mobile ad hoc networks," *Proceedings of the 6th Annual International Conference on Mobile Computing and Networking*, Boston, Massachusetts, United States, pp. 255–265.

[8] D. B. Johnson and D. A. Maltz, "Dynamic source routing in ad hoc wireless networks," in *Mobile Computing*, (T. Imielinski and H. Korth, Editors), Kluwer Academic Publishers, 1996, chap. 5, pp. 153–181.

[9] Y. Zhang, W. Lee, and Y.-A. Huang, "Intrusion detection techniques for mobile wireless networks," *ACM J. Wireless Networks*, vol. 9, no. 5, Sep. 2003, pp. 545–556.

[10] Y. Okazaki, I. Sato, and S. Goto, "A new intrusion detection method based on process profiling," *Symposium on Applications and the Internet*, 28 Jan.–1 Feb. 2002, pp. 82–90.

[11] P. Albers, O. Camp, J.-M. Percher, *et al.*, "Security in ad hoc networks: a general intrusion detection architecture enhancing trust based approaches," *1st International Workshop on Wireless Information Systems, 4th International Conference on Enterprise Information Systems*, Ciudad Real, 3–6 Apr. 2002.

[12] A. B. Smith, "An examination of an intrusion detection architecture for wireless ad hoc networks," *5th National Colloquium for Information System Security Education*, May 2001.

[13] O. Kachirski and R. Guha, "Intrusion detection using mobile agents in wireless ad hoc networks," *IEEE Workshop on Knowledge Media Networking*, 10–12 Jul. 2002, pp. 153–158.

[14] A. Mishra and K. M. Nadkarni, "Security in wireless ad hoc networks – a survey," in *The Handbook of Ad Hoc Wireless Networks* (M. Ilyas, Editor), CRC Press, 2002, chap. 30.

[15] A. Mishra, K. Nadkarni, and A. Patcha, "Intrusion detection in wireless ad hoc networks," *IEEE Wireless Communications*, vol. 11, no. 1, Feb. 2004, pp. 48–60.

[16] K. Nadkarni and A. Mishra, "A novel intrusion detection scheme for wireless ad hoc networks," *IEEE WCNC '04*, vol. 2, Mar. 2004, p. 831.

6

Quality of service

Wireless mobile ad hoc networks consist of mobile nodes interconnected by wireless multi-hop communication paths. Unlike conventional wireless networks, ad hoc networks have no fixed network infrastructure or administrative support. The topology of such networks changes dynamically as mobile nodes join or depart the network or radio links between nodes become unusable. Supporting appropriate quality of service for mobile ad hoc networks is a complex and difficult issue because of the dynamic nature of the network topology and generally imprecise network state information, and has become an intensely active area of research in the last few years. This chapter presents the basic concepts of quality of service support in ad hoc networks for unicast communication, reviews the major areas of current research and results, and addresses some new issues. The focus is on routing issues associated with quality of service support. The chapter concludes with some observations on areas for further investigation.

6.1 Introduction

Mobile ad hoc networks offer unique benefits and versatility for certain environments and certain applications. Since a fixed infrastructure, including base stations, is not necessary, they can be created and used "any time, anywhere." Second, such networks could be intrinsically fault-resilient, for they do not operate under the limitations of a fixed topology. Indeed, since all nodes are allowed to be mobile, the composition of such networks is necessarily time varying. Addition and deletion of nodes occur only by interactions with other nodes; no other agency is involved. Such perceived advantages elicited immediate interest in the early days among military, police, and rescue agencies in the use of such networks, especially under disorganized or hostile environments, including isolated scenes of natural disaster and armed conflict.

In recent days, home or small-office networking and collaborative computing with laptop computers in a small area (e.g., a conference or classroom, single building, convention center, etc.) have emerged as other major areas of application. In addition, people have recognized from the beginning that ad hoc networking has obvious potential use in all the traditional areas of interest for mobile computing.

Mobile ad hoc networks are increasingly being considered for complex multimedia applications, where various *quality of service* (QoS) attributes for these applications must be satisfied as a set of predetermined service requirements. As a minimum, the QoS issues pertaining to delay and bandwidth management become of paramount interest. In addition, because of the use of ad hoc networks for military or police use, and increasingly common commercial applications, various security issues also need to be addressed for such applications, which I do not address here. Cost-effective resolution of these issues at appropriate levels is essential for widespread general use of ad hoc networking.

Mobile ad hoc networking emerged from studies on extending traditional Internet services to the wireless mobile environment. All current works, as well as this book, consider ad hoc networks as a wireless extension to the Internet based on the ubiquitous IP networking mechanisms and protocols. Today's Internet possesses an essentially static infrastructure where network elements are interconnected over traditional wire-line technology, and these elements, especially the elements providing the routing or switching functions, do not move. In a mobile ad hoc network, by definition, all the network elements move. As a result, numerous more stringent challenges must be overcome to realize the practical benefits of ad hoc networking. These include effective routing, medium (or channel) access, mobility management, power management, and security issues, all of which affect the quality of the service experienced by the user.

The absence of a fixed infrastructure for ad hoc networks means that the nodes communicate directly with one another in a peer-to-peer fashion. The mobility of these nodes imposes limitations on their power capacity, and hence, on their transmission range; indeed, these nodes must often satisfy stringent weight limitations for portability. Mobile hosts are no longer just end systems; to relay packets generated by other nodes, each node must be able to function as a router as well. As the nodes move in and out of range with respect to other nodes, including those that are operating as routers, the resulting topology changes must somehow be communicated to all other nodes as appropriate. In accommodating the communication needs of the user applications, the limited bandwidth of wireless channels *and* their generally hostile transmission characteristics impose additional constraints on how much administrative and control information may be exchanged, and how often. Ensuring effective routing is one of the great challenges for ad hoc networking.

The lack of fixed base stations in ad hoc networks means that there is no dedicated agency for managing the channel resources for the network nodes. Instead, carefully designed distributed medium access techniques must be used for channel resources, and, hence, there must be mechanisms available to recover efficiently from the inevitable packet collisions. Traditional carrier-sensing techniques cannot be used, and the hidden terminal problem may significantly diminish the transmission efficiency. Although I do not expand on this issue further, an effectively designed protocol for medium access control (MAC) is essential for the quest for QoS; see, for example, [1] and the references cited therein for additional information.

All the challenges enumerated above are potential sources of service impairment in ad hoc networks and, hence, may degrade the "quality of service" seen by the users. As of now, the Internet has only supported the "best effort" service; best effort in the sense that it will do its best to transport the user packets to their intended destination, although without any guarantee. Quality-of-service support is recognized as a challenging issue for the Internet, and a vast amount of research on this issue has appeared in the literature during the last decade or so [2]. With the Internet as the basic model, the ad hoc networks have been initially considered only for "best effort" services as well, especially given their peculiar challenges when compared with traditional wire-line or even conventional wireless networks. Indeed, just as the QoS accomplishments for wired networks, such as the Internet, cannot even be directly extended to the wireless environment, QoS issues become even more formidable for mobile ad hoc networks. Happily, during the last few years quality of service for ad hoc networks has emerged as an active and fertile research topic of a growing number of researchers and major advances are expected in the next few years. See [3] for a comprehensive review of the state-of-the-art on QoS routing in ad hoc networks, circa 1999. The URLs of [4] are good sources of more up-to-date information in this area.

Performance of these various protocols under "field" conditions is, of course, the final determinant of their efficacy and applicability. Relative comparisons of computational and communication complexities of various routing protocols for ad hoc networks have appeared in the past, e.g., [3, 5, 6, 7], providing the foundation for more application-oriented assessment of their effectiveness. On the other hand, the performance studies have started to appear only recently, e.g., [8]. The mathematical analysis of ad hoc networks, even under the simplest assumptions about the dynamics of the topology changes and the traffic processes, poses formidable challenges, and even their simulation is considerably more difficult than their static counterpart. Performance studies of ad hoc networks with QoS constraints remain an open area of research.

I also observe that secure QoS routing remains essentially an open area of research for ad hoc networks, which I do not address here. See [9] for a survey of the security issues for mobile ad hoc networks circa 2001.

My discussion is limited to unicast communication only; multicasting adds additional layers of complexity to the problems of unicast communication and requires its own separate survey. See [3, 4] for additional information on the QoS issues associated with multicast routing.

The organization of the rest of the chapter is as follows. Section 6.2 introduces some networking concepts pertinent to routing and QoS. The general issue of routing in mobile ad hoc networks is reviewed in Section 6.3. Section 6.4 addresses QoS routing issues for ad hoc networks and its current state of research. Finally, Section 6.5 presents concluding remarks and suggests directions for future research. For a general survey of all the issues mentioned here, see [10].

6.2 Routing in mobile ad hoc networks

All routing protocols for ad hoc networks need to perform a set of basic functions in the form of route identification and route reconfiguration. For communication to be possible, at least one route (i.e., a loop-free path) must exist between any pair of nodes. Route identification functions, as the name suggests, identify a route between a pair of nodes as a prerequisite to communication. Route reconfiguration functions are invoked to recover from the effects of undesirable events such as host or link failures of various kinds, and traffic congestions appearing within a subnetwork. Evidently, recognition of changes in the network topology and the topology update functions constitute an indispensable subset of the route reconfiguration functions. A separate category of resource management functions is also considered, to ensure that all the network resources are available, to the extent possible, in support of some special objectives such as those associated with QoS or security. Different authors use different classification schemes for these basic routing functions.

Routing in ad hoc networks, as in their wired counterparts, has traditionally used knowledge of the instantaneous connectivity of the network with emphasis on the state of the links. This is the so-called topology-based approach [6]; the associated routing protocols can generally be classified into three categories; periodic (also called proactive or table-driven), on-demand (also called reactive), and hybrid protocols.

Networks using periodic protocols attempt to maintain the knowledge of every current route to every other node by periodically exchanging routing information, regardless of whether the routes are being used for carrying packets. Each node maintains the necessary routing information and the nodes are responsible for propagating topology updates in response to instantaneous connectivity changes in the network. Examples of such protocols include those based on destination-sequenced distance-vector (DSDV) routing [11] and its derivatives, among many others. As a class, these protocols tend to suffer from wasted bandwidth due to the large control overhead in maintaining unused routes, especially during frequent changes in network topology, although some of the newer link-state routing protocols mentioned in [10] present approaches for reducing the overhead.

The on-demand protocols, in contrast to periodic protocols, create routes only when necessary for carrying traffic. As a result, a route discovery process is a prerequisite to establishing communication between any two nodes, and a route is maintained as long as the communication continues. Examples of on-demand protocols include dynamic source routing (DSR) [12], ad hoc on-demand distance vector (AODV) routing [13], and many others cited in [10].

The on-demand protocols also tend to generate large overheads and suffer loss of packets in transit as the topology changes become more frequent. However, in general, these protocols perform better than their periodic counterparts, especially when topology changes are infrequent, e.g., see [8]. A survey of periodic and on-demand routing protocols including their relative time and communication complexities, circa 1999, is presented in [7].

The hybrid approach combines both aspects of periodic and on-demand routing. For example, the zone routing protocol (ZRP) [14] allows the use of a periodic routing protocol within a local zone, while an on-demand routing scheme is used globally. Thus, at least at the level of inter-zone routing, if the topology changes are not too frequent, the benefits of on-demand routing are available. The performance of the zone routing protocol clearly depends on the organization of the zones within the network and the traffic patterns within the zones, neither of which is particularly predictable under all circumstances.

A different approach, called location-based (or position-based) routing, aims to reduce some of the drawbacks of topology-based routing. A recent survey, including comparative information on the time and communication complexities of various protocols of this category, is presented in [6]. In addition to topology-based information, these protocols also use information about the physical location of the mobile hosts. The Global Positioning System (GPS) is used often by the nodes to determine their respective positions.

A distinguishing characteristic of location-based protocols is that to forward packets, a node only requires its own position, that of the destination (obviously), and those of its adjacent (one-hop) neighbors. A transmitting node uses a location service to determine the location of the destination, and includes this location information as part of the destination address in its messages. Routes do not need to be established or maintained explicitly; thus there is no need to store routing tables at the nodes, and no need for routing table updates. Adjacent nodes are identified typically by broadcasting limited-range beaconing messages and various time-stamping mechanisms. The beaconing message includes distance limits; a receiving node discards the message if its location lies beyond the distance limit.

Availability of accurate location information at each node is essential for location-based routing to work, which, in turn, requires timely and reliable location updates as nodes change their locations. One or more nodes, designated to act as location servers, coordinate these location service functions, which are necessarily decentralized because of the mobility of the nodes. A large part of the ongoing research, as the references cited above show, is focused on designing efficient location services.

Performance studies on location-based routing, similar to those reported above for topology-based routing, are yet to appear in the literature. At a first glance, it appears reasonable to expect that QoS objectives would be easier to meet by avoiding routing updates. More work is needed to confirm this expectation. Finally, I am not aware of any reported work on security issues for such protocols.

A high-level overview of routing with QoS constraints follows next as a prologue to the more detailed discussion of these issues for ad hoc networks.

6.3 Routing with quality of service constraints

The RFC 2386 standard [15] characterizes QoS as a set of service requirements to be met by the network while transporting a packet stream from the source to the destination. Intrinsic to the notion of QoS is an agreement or a guarantee by the network to provide a set of measurable pre-specified service performance constraints for the user in terms of end-to-end delay, delay variance (jitter), available bandwidth, probability of packet loss, etc. The cost of transport and total network throughput may also be included as parameters. Obviously, enough network resources must be available during the service invocation to honor the guarantee. The first essential task is to find a suitable route through the network between the source and destination that will have the necessary resources available to meet the QoS constraints for the desired service. The

Figure 6.1 A flow: QoS is meaningful only for a flow between a specific source–destination pair

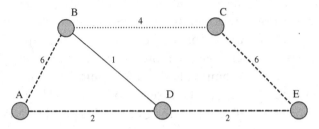

Figure 6.2 Links with fixed bandwidth in a network

task of resource (request, identification) reservation is the other indispensable ingredient of QoS. By QoS routing, I mean both these tasks together. The Internet of today operates in a connectionless and stateless mode. The network of routers is not aware of any association between the source and destination except on a per-packet basis. Each packet is routed individually without any information about the state of the flow of packets between the source and destination. On the other hand, QoS is meaningful only for a flow of packets between the source and destination, and thus depends on the notion of a logical association, or logical connection, between them for the duration of the flow, as represented in Fig. 6.1. Second, the network must guarantee the availability of a set of resources associated with the flow. Consequently, appropriate routers must remain aware of the logical connection and the state of the flow to ensure that adequate network resources, such as link bandwidth, nodal buffers, processing power, etc., are available for the duration of the logical connection, and their underlying routes. Quality of service guarantees can be attained only with appropriate resource reservation techniques. The most important element among them is QoS routing, i.e., the process of choosing the routes to be used by the flow of packets of a logical connection in attaining the pre-established QoS guarantee.

Consider Fig. 6.2, where the numbers next to the links represent their respective bandwidth, say in Mbps. To minimize delay and for better use of network resources, minimizing the number of intermediate hops is one of the principal objectives in determining suitable routes. However, suppose that the packet flow from A to E requires a bandwidth guarantee of 3 Mbps. Quality of service routing will then select the route A–B–C–E over the route A–E, since the latter is unable to meet the bandwidth need although it has fewer hops. The only other alternative, A–B–D–E, will also be rejected for failing to meet the bandwidth need.

Quality of service routing offers serious challenges even for the static environment of today's Internet. Different service types, e.g., voice, live video, and document transfer, have significantly different objectives for delay, bandwidth, and packet loss. Determining the QoS capability of candidate links is not simple for such scenarios (for multicast services, the difficulties are even greater). It has already been noted that the route computation cannot take "too long." Consequently, the computational and communication complexities of route selection criteria must also be taken into account. The presence of more than one QoS constraint often makes the QoS routing problem NP-complete [3]. Sub-optimal algorithms such as sequential filtering is often used, especially for large networks, where an optimal path based on a single primary metric (e.g., bandwidth) is selected first, and a subset of them are eliminated by optimizing over the secondary metric (e.g., delay), and so on, until all the metrics have been taken into account (a random selection is made if there is still more than one choice after considering network throughput as the last metric). All else remaining the same, the same route is used for all the packets in the flow as long as the QoS constraints are satisfied.

Candidate routes for a flow with specific QoS objectives are determined by using various QoS metrics associated with its constituent nodes and links. These metrics collectively characterize the state of the nodes and links. A typical link-state is an ordered tuple of its specific QoS metrics of interest, and is usually represented as follows:

$$\text{link-state} = <\text{bandwidth}, \text{propagation delay}, \text{cost}>,$$

where bandwidth is the maximum residual bandwidth that the link can support, and "cost" here is used as a generic catch-all for other parameters such as packet loss statistics, service class (if multiple service classes are to be supported, each with its own QoS requirements), etc. "Cost" often does represent a single number. Observe also that each link is assigned a unique direction in terms of the source and sink nodes. The state of a node is likewise characterized as an ordered tuple of typical QoS metric as follows:

$$\text{node-state} = <\text{CPU bandwidth capacity}, \text{delay distribution}, \text{cost}>,$$

where the CPU bandwidth is the minimum rate at which the node can place data into the link, delay distribution at a minimum includes the mean and the variance of the queueing delay, and "cost" is, again, a generic term for many other parameters that need to be considered for different service classes for different traffic types, including service classes with multiple priorities. Frequently, the node-state is incorporated into the state of each of the links incident on it, as in this chapter. In such cases, we only have (augmented) link-states, where the link bandwidth is now the minimum of the residual link bandwidth and the CPU bandwidth of its source node, and the delay is the (random) sum of the link propagation and node queueing delays. Finally, the cost is determined appropriately by considering its component metrics.

Accurate location information has to be included as part of the local state information, if location-based routing is to be considered.

The state of a route, such as that shown in Fig. 6.2, then follows immediately as the appropriate numerical operations of the various components of the (augmented) states of its constituent links. The bandwidth of a route is the minimum of that of its components, the delay is the (random) sum of all link delays, and the cost is either the sum (if it is an additive quantity), or another appropriate deterministic or stochastic numerical operation of all the component costs. For a given flow, a feasible route is one with sufficient available resources to satisfy the QoS requirements. It is evident then that the QoS routing problem is a constrained combinatorial (graph) optimization problem, and it is solved as such.

Once a route has been selected for a specific flow, the necessary resources, e.g., bandwidth, buffer space in routers, etc., must be reserved for the flow. These resources will not be available to other flows until the end of this particular flow. Consequently, the amount of remaining network resources available to accommodate the QoS requests of other flows will have to be recalculated, and propagated to all other pertinent nodes as part of the topology update information.

Quality of service routing being dependent on the accurate availability of the current network state, I briefly consider the nature of such information. The first is the local state information maintained at each node, which includes every pertinent component (for a given flow) of the node-state, as well as the link-state for each of its outgoing links. The totality of the local state information for all nodes constitutes the global state of the network, which is also maintained at each node. The instantaneous network connectivity is part of the global state information. While the local state information may be assumed to be always available at any particular node, the global state information is constructed by exchanging the local state information for every node among all the network nodes at appropriate moments. The process of updating the global state information, also loosely called topology updating, may significantly affect the QoS performance of the network, as has been mentioned before.

The global state may be updated by broadcasting the local state of each node to every other node (link-state protocol), or by periodically exchanging

Quality of service

Table 6.1

	Bandwidth entry at Node A			
Destination	B	C	D	E
Bandwidth	6	4	2	4
Adjacent node	B	B	D	B

suitable "distance vector" information among the adjacent nodes only (distance-vector protocol). A distance vector is usually maintained as a table in each node with an entry for each QoS metric. Given a node K, for each QoS metric, the corresponding entry is a triplet consists of the following:

- Address of a destination node for every possible destination;
- Best attainable value of the metric over the best route from K to the destination; and
- Address of the node immediately adjacent (next hop) to K on the best route with respect to the value of the metric.

Consider node A in the example of Fig. 6.2. The bandwidth entry for the distance-vector at the node A will have a representation as given in Table 6.1.

For destination B, the available routes are $\langle A \rightarrow B \rangle$, $\langle A \rightarrow D \rightarrow B \rangle$, and $\langle A \rightarrow D \rightarrow E \rightarrow C \rightarrow B \rangle$. Likewise for E, the available routes are $\langle A \rightarrow D \rightarrow E \rangle$, $\langle A \rightarrow B \rightarrow D \rightarrow E \rangle$, $\langle A \rightarrow B \rightarrow C \rightarrow E \rangle$, and $\langle A \rightarrow D \rightarrow B \rightarrow C \rightarrow E \rangle$, and so on.

Since the topology updates throughout the network cannot happen instantaneously, the global state information may only be an approximation of the true current network state. For ad hoc networks with highly mobile nodes, the global state information may never be accurate.

Three distinct route-finding techniques are used to determine an optimal path satisfying the QoS constraints. These are source routing, distributed routing, and hierarchical routing. In source routing, a feasible route is computed locally at the source node using the locally stored global state information, and then all other nodes along this feasible route are notified by the source of their adjacent preceding and successor nodes. A link-state protocol is almost always used to update the global state at every node. The state update is done by using either a distance-vector (most common) or a link-state protocol. In distributed or hop-by-hop routing, the source and other nodes are involved in the path computation by identifying the adjacent router to which a node must forward the packet associated with the flow. Practical considerations for large networks with many nodes and high connectivity sometimes compel the use of so-called aggregated global state information, by first partitioning the network into a hierarchical cluster of some form, and then only considering the suitable state information associated with these clusters. Such information is, necessarily, a partial representation of the true global state. Hierarchical routing, as the name suggests, uses the aggregated partial global state information to determine a feasible path using source routing where the intermediate nodes are actually logical nodes representing a cluster; for more details see [16]. Flooding is not an option for QoS routing, except for broadcasting control packets under appropriate circumstances, e.g., for beaconing, or at the start of a route discovery process. See [7] for a comparative discussion of the advantages and disadvantages of various algorithms associated with each of these three approaches.

One may reasonably expect that all packet exchanges will not be treated with equal priority in a QoS network. The exchange of control packets should receive higher priority than user data packets in a network designed for QoS. Indeed, except for instances of "thin" low-traffic (relative to the network capacity) networks, control packets should receive pre-emptive priorities over user data packets. Second, the QoS policy may allow different priorities to exist even among different flows of user packets. Clearly, in accommodating packets with pre-emptive priorities, the network may not be able to preserve the QoS guarantee for ordinary flows. Appropriate admission control policies could also offer additional benefits. Indeed, QoS routing allowing pre-emption and admission control policies is an open area for further research.

Handling of user data with multiple priorities presents potential security threats as well. When a user requests QoS with a certain priority, the network first needs to authenticate such a request by exchanging appropriate control packets. Too many authentication requests, in themselves, may degrade the operational performance of a large QoS network. Next, the network must find a route with the requested QoS for a higher priority against all other flows with lesser priority, even if they are allocated identical QoS parameters in all other respects. In heavy traffic situations, guaranteeing QoS for lesser priority traffic may be extremely difficult or impossible. The development of QoS routing policies, algorithms, and protocols for handling user data with multiple priorities is also an open area.

Similar challenges exist in designing QoS routing schemes supporting multiple service classes. For additional details, see [3].

My discussion, up to this point, has been limited to unicast routing. The essential problem here is to find a feasible path from a source node to a single destination node that satisfies a set of QoS constraints, and possibly some other additional optimization criteria such as minimum cost and maximum network throughput. The multicast routing problem, on the other hand, is distinguished by more than one destination node, where the objective is to find not a single path, but a feasible tree rooted at the source. Each path from the source to one of the destination nodes in the tree is required to satisfy the specified set of QoS constraints, together with additional optimization criteria, if any, simultaneously. As observed in [3], many of the associated optimization problems are NP-complete. In particular, most categories of the general constrained combinatorial optimization problems for graphs are known to be NP-complete. I do not address the topic of multicast routing in this chapter.

This section has presented only a broad-brush overview of QoS routing. Many issues such as the effect of imperfect knowledge of network state information on routing, and hierarchical aggregation of routing information for scalability, etc., have not been mentioned above. All these issues profoundly affect the QoS in ad hoc wireless networks, and are considered in the next section.

6.4 Quality of service routing in ad hoc networks

The basic concepts of the QoS routing discussed in the previous section constitute the foundation for QoS routing for ad hoc networks. This discussion is based on topology-based routing; observations pertinent to location-based routing are added as appropriate. I assume that each node carries a unique identity recognizable within the network. Following [3], I assume the existence of all necessary basic capabilities, such as suitable protocols for medium access control and resource reservation, resource tracking, state updates, etc.

Each node periodically broadcasts a beacon packet identifying it and its pertinent QoS characteristics, thus allowing each node to learn of its adjacent neighbors (i.e., those with which it can communicate directly). The beaconing mechanism, regardless of whether it is topology-based or location-based, lies at the heart of ad hoc networking, for otherwise a node will not even know its adjacent neighbors, which change dynamically in such networks. The knowledge of adjacent neighbors, of course, is indispensable for routing.

A principal objective of network engineering, as emphasized earlier, is the minimization of routing updates, for such updates consume network bandwidth and router CPU capacity. Second, frequently changing routes could increase the delay jitter experienced by the users. This objective is extremely difficult to attain in wireless networks because of the involuntary network state changes as nodes join in or depart, traffic loads vary, and link quality swings dramatically. To accommodate real time traffic needs, such as those of voice or live video, both the overall delay and the delay variance must be kept within a certain bound. This is accomplished primarily by minimizing, as far as possible, the number of hops, or the intermediate routers, in the path. With potentially unpredictable topology changes in an ad hoc network, this last objective is difficult to attain.

Combinatorial stability, therefore, is a critical consideration for QoS in an ad hoc network. Combinatorial stability follows directly when the geographical distribution of the mobile nodes does not change much relative to one another during the time interval of interest. Such is the case, for example, for the Internet, and in a classroom setting for communication among laptop computers as ad hoc nodes. The routes among network nodes in such cases will change little or not at all. There are other cases, for example, in rescue operations, refugee migrations, etc., where the route updates do occur during the intervals of interest, but not sufficiently frequently to violate the limits of combinatorial stability. In such cases, it is possible that the topology updating takes long enough so that by following the now unacceptable characteristics of the last used route, the QoS guarantees cannot be met. Indeed, the old route may even cease to exist during the topology update. This is entirely possible for geographically dispersed networks with a large number of nodes and sparse connectivity, where each route consists of many intermediate nodes, like a string of beads.

The topology of an ad hoc network may be combinatorially just right so that the QoS guarantees are maintained during any topology updating. Observe that it is not just the connectivity that affects the QoS, but the availability of enough resources along the previous and the new routes during and after the transition is equally essential. We call an ad hoc network QoS-robust with respect to a specific set of QoS guarantees only if such guarantees are maintained regardless of the topology updates that may occur within the network; guaranteeing QoS-robustness under all circumstances is possible only with unlimited resources. More narrowly, we call such a network QoS-preserving if it can continue to maintain the QoS guarantees during the interval from the end of a successful topology update until the occurrence of the next topology change event. A QoS-robust ad hoc network is, by definition, QoS-preserving; the converse is obviously false.

A mobile node may lose connectivity with the rest of the network simply because it has wandered too far off, or its power reserve has dropped below a critical threshold. Since the network cannot control the occurrence of such events, we must exclude them in considering the QoS guarantees. A topology update occurs when a new node joins the network or an existing node is detected to have become unavailable with respect to a particular flow. One naturally expects that such topology updates should not affect the QoS for the rest of the nodes as long as the topology of the rest of the network (as a subnetwork) remains unchanged. So far, with the exception of [3], little has appeared on the preservation of QoS guarantees under various failure conditions in ad hoc networks as a specific area of study.

Two routing techniques are considered in [3], both limited to combinatorially stable, QoS-preserving networks. One is based on the availability of only local state information, and the other assumes a possibly inaccurate knowledge of global states. When an existing feasible route becomes unavailable, a new feasible path is determined, and the flow is rerouted to the new feasible path. During the interval immediately following the disappearance of the existing path and the establishment of the new route, data packets are sent as best-effort traffic.

For QoS routing using only the local state information, [3] introduces two different distributed routing algorithms, the so-called source initiated routing and destination-initiated routing. Both use only the local state information stored at each node. Both rely on the use of probe packets with appropriate nodal identity and QoS information in identifying a feasible route with the desired QoS characteristics. The source and the intermediate routers use a form of flooding to send the probe packets. Various mechanisms are considered in [3] to mitigate the penalties of flooding and to minimize the number of probe packets to be used, and the advantages of destination-initiated routing over the other methods established under certain conditions.

Pre-established network policies should determine the steps to be taken, in case no feasible route could be found during the route establishment phase. The service request may be rejected, and the node blocked, or the network may negotiate for a service with lower QoS by exchanging control packets using best-effort routing, assuming that such alternative QoS is available. Such considerations offer opportunities for further research.

Efficient source-initiated routing results from a number of innovative techniques introduced in [3]. Avoiding unnecessary probes by noting their respective sources is one. The second is the novel concept of local multicast, which limits the broadcast of probes to only an appropriate subset of the adjacent nodes. The third relies on caching the distance information by counting the number of hops traversed by the probe up to that point. By maintaining, at each node, the relevant state information of all its n-hop neighbors, a route to any other node can be determined by using only local information. It is evident that the location-based routing techniques mentioned in [6] perform similar functions without the need for route updates, and offer opportunities for potential improvements in efficiency.

The destination-initiated routing approach of [3] actually relies on the best available estimate of the distance between the source and destination. Here the destination node identifies a feasible route by sending probe packets towards the source on the basis of restricted flooding. Of course, it is the source that initiates a flow by sending a control message to the destination with the necessary QoS information by using one of the many best effort routing algorithms mentioned in Section 6.3. The control message counts the number of hops it traverses while following the best effort route as an estimate (upper bound) of the distance between the source and the destination. This hop count is used at the destination node to limit the flooding range for its probe packets back toward the source. More precise location information used in location-based routing should result in more accurate restriction on the flooding range, thus offering opportunities for greater efficiency. The techniques based on imprecise knowledge of global states in [3] use the notion of ticket based probing for identifying a feasible route. Each probe from the source towards the destination carries at least one ticket to control the number of alternate paths to be searched, thus minimizing the routing overhead. The lower the likelihood of finding a route with the desired QoS requirements, the larger is the number of tickets carried by the probe. The probes are attempted to be sent along links, the QoS characteristics of which are relatively constant (or slowly varying) in time. The basic routing mechanism is distributed or hop-by-hop; in [3], the information for multiple feasible routes is stored in the probes, instead of within the intermediate routers.

Several mechanisms are considered in [3] for QoS-preserving QoS routing by detecting broken routes and then either repairing the broken route or by rerouting the flow on an alternate route with the desired QoS. The use of redundant routes of various kinds further reduces the likelihood of QoS violation. A broken route is detected by a node on the route using a mechanism similar to the beaconing protocols for detecting adjacent neighbors. When a node detects a broken route, it sends a "route failure" message back to the source. After receiving the "route failure" message, the source switches the flow over to an alternate route, as discussed below, and sends a "resource release" message along the original route so that all nodes on the route receiving this message can release all QoS resources previously reserved for the flow. Obviously, the "resource release" message will not reach those nodes on the now broken route that are no longer reachable from the source. Even then, their resources will not remain associated with the now-rerouted flow indefinitely, by using the following "time-out" mechanism. The existence of the QoS route between a source–destination pair needs to be reaffirmed periodically when routing with imprecise information by sending suitably constructed control packets, called refresher packets in [3], from the destination back to the source. When an intermediate node receives the refresher packet, it resets the "refresher timer" and sends the refresher packet to the adjacent node upstream. The receiving node always sends an acknowledgment back to the sending node. If such a refresher packet fails to arrive within a predetermined

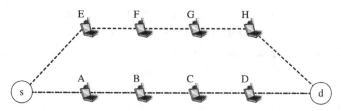

Figure 6.3 Alternate routing

timeout interval, the QoS route is declared unavailable and the associated resources released. Likewise, if a node expecting to receive an acknowledgment to the refresher packet does not receive it before the "refresher acknowledgment timeout" expires, it releases all of its own resources associated with the particular QoS route. This also accommodates the failure to reach various unavailability notifications to their intended recipients using additional timeout mechanisms, such as timeouts on timeout messages.

When the source receives the notification of route unavailability, it seeks an alternate route with the same QoS characteristics, as shown in Fig. 6.3.[1] The unusable route is shown by a dashed line, and the new alternate route is shown by a solid line. If such a route can be found, the flow is rerouted to it after the necessary route updates among the pertinent nodes.

Several redundant routing mechanisms are also considered in [3] for minimizing the likelihood of QoS violation owing to route failures. At the highest level of redundancy, multiple alternate routes with the same QoS guarantee are established for the flow, and are used simultaneously.

The alternate routes should preferably be disjoint,[2] although this may not always be possible, as shown in Fig. 6.4. At the next lower level of redundancy, the routes and the associated resources are reserved and rank ordered, but not used unless the primary route fails, or the first choice for the alternate route fails while the primary route is unavailable, and so forth. When not in use for the QoS-guaranteed flow, the alternate route is used to carry best-effort packets. At the lowest level of redundancy, only the route is identified; no resource is reserved. When the primary path fails, the alternate paths are checked to determine whether the necessary resources are still available. An explicit discovery process for rerouting is initiated if none of the alternate

---

[1] Recall that in developing various routing and other algorithms for ad hoc networks, minimizing power consumption has been explicitly investigated by many researchers. Minimization of power consumption and QoS support do not appear to be mutually consistent objectives at the current "state of the art."

[2] Two routes are disjoint if, and only if, the source and destination nodes are only common nodes in the routes.

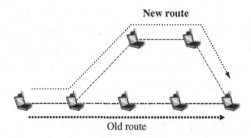

Figure 6.4 Redundant routing: all routes are not disjoint

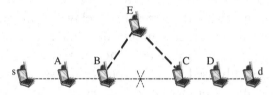

Figure 6.5 Route repair

routes are found to be able to support the desired QoS. In all cases, the duplicate packets are discarded at the destination.

Variations of the above approach are possible; one is where an attempt is made to repair the route solely on the basis of local adjacency information, instead of switching over to a new route. Instead of the source, if node B in Fig. 6.5 determines that the link to C is broken, it does not send a route failure message back to the source. Instead, B attempts to repair the path as follows. Using the beaconing mechanism, B sends a "repair request" message to its adjacent neighbors, querying whether any of these other nodes may be able to offer at least the same QoS support as C. An adjacent neighbor E will send an affirmative response only if it is also an adjacent neighbor to C with a link, and if it has adequate residual resources. If node E sends an affirmative response, B will add the link as an element to the QoS route from s to d, and send a "path repair" message to E. After receiving the path repair message, E will dedicate the necessary resource to the QoS flow and update its own route information. The new information will become part of the regular topology update as required.

None of these scenarios explicitly considers location-based routing. The recently introduced notion of predictive location-based QoS routing [17] is an attempt to exploit explicitly the potential advantages of location-based routing in connection with alternate routing. In this approach, the route failures are predicted beforehand, and new routes are determined using these predictions before the current route becomes unavailable. In principle, the

QoS flow can be transferred on to one of these predicted alternate routes without any packet loss. The nodes in the network use flooding for state updates consisting of their respective locations and resource availability. The location updates are used to predict the future locations of the nodes, and alternate routes are determined using the predicted locations such that the connectivity between the source and destination on the new route will be preserved. However, this approach does not reserve resources for the alternate routes and, as is to be expected, does not promise any hard QoS guarantees. The availability of resources on all active routes is appropriately monitored against the QoS objectives to ensure the occurrence of the necessary route switching when the available resources drop below acceptable thresholds.

Admission control policies, common in wire-line networks, offer opportunities for preserving existing QoS in an ad hoc network. In wire-line networks, these are most frequently used with multiple QoS classes having different QoS attributes and priorities. For mobile ad hoc networks, the "best effort" traffic is the most natural QoS class, say, service class 0. Multimedia traffic, with and without live video, may be assigned their own QoS classes. When a new node attempts to find a QoS route for service with a QoS class, the route discovery will fail if there is no feasible QoS route available at that moment to accommodate. An admission control policy for ad hoc networks should answer questions such as whether the requesting node could negotiate with the destination for a "lower QoS;" if such negotiations are allowed, then the policy will also specify how many, how often, and for how many different "values." The default option, as considered in [3], could be to switch to "best effort" service only. More complex options will necessitate addressing issues such as whether to pre-engineer for alternate QoS options, or use adaptive negotiations. A robust admission control policy will take into account the effect of additional control traffic on the QoS capacity of the network.

I have mentioned earlier the option of assigning control packets with pre-emptive priorities over other "data" packets as part of strengthening the QoS support. One may reasonably expect that all packet exchanges will not be treated with equal priority in a QoS network. Likewise, different levels of QoS may include different priorities for different flows. The routing protocol must find a route with the requested QoS for a higher priority against all other flows with lesser priority, even if they are allocated identical QoS parameters in all other respects. In heavy traffic situations, guaranteeing QoS for lesser priority traffic may be difficult or even impossible. In such cases, the admission control policy needs to address whether the QoS guarantee for flows could be preserved, and what actions to take in case they are not. The development of QoS routing with admission control policies, algorithms, and protocols, with or without control packet priorities or multiple levels of priorities for user data, is an area for further research.

I have repeatedly noted that all the policies, protocols, and algorithms in an ad hoc network with QoS support must be QoS-preserving. How badly do the rapid topology changes militate against the QoS guarantees? Let τ_c and T_u denote the interval between two consecutive topology change events and the time it takes to detect the change, complete the calculation, and propagate the topology updates resulting from the last topology change, respectively, to all pertinent nodes.[3] Recall that an ad hoc network is combinatorially stable only if $T_u < \tau_c$. If the computed feasible route ceases to exist during the corresponding topology update, the QoS guarantee becomes meaningless. Maintaining bounds on delay jitter may also become impractical even in a combinatorially stable network if τ_c remains "close" to T_u. It may be necessary to investigate more rigorous criteria for different degrees of combinatorial stability and for different QoS constraints. Since combinatorial stability is governed by random processes arising from random changes in the topology and link traffic intensity, making the network QoS-robust for a particular flow and its associated QoS constraints is clearly impossible as a deterministic objective for an arbitrary ad hoc network. This is why no QoS routing algorithm offers hard QoS guarantees now, nor are they likely to in the future. Any such guarantee could at best only be statistical in nature, where QoS robustness is specified as a probability bound for QoS violation during a topology update, the duration of which does not exceed a fixed upper bound. This is why any performance study of ad hoc networks with QoS support is meaningful only for combinatorially stable networks. This is also why one assumes that the connectivity between a node and the rest of the network is never lost because of low battery power, or because a mobile node has wandered far enough away. The smaller the value of T_u, the smaller is the probability of QoS violation, assuming that resources remain available for use whenever necessary. Redundant routing, as mentioned earlier, clearly could help accomplish both. It is obvious that QoS is a realistic goal to pursue for static ad hoc networks, e.g., a classroom setting. It is equally obvious that considerable additional work is necessary to understand better both the specific conditions and extents under which various QoS objectives could be satisfied for the dynamic ad hoc networks in the real world. In this latter context, the use of admission control, multiple classes of service with possibly different priorities, pre-emptive priority of control messages, and segregation of dedicated resources for QoS-robust ad hoc networking offer promising areas of investigation.

---

[3] In practice, these are random variables.

6.5 Conclusion and further reading

I have attempted a terse introduction to the new, but rapidly growing, area of research on guaranteeing QoS in ad hoc mobile wireless networks. The issues are challenging, many of the underlying algorithmic problems are currently perceived as generally intractable (NP-complete), and opportunities exist for creating more effective heuristics. The issues are complicated by the lack of sufficiently accurate knowledge, both instantaneous and predictive, of the states of the network, e.g., the quality of the radio links, and availability of routers and their resources. The successful QoS routing includes the necessary knowledge of the network state, and algorithms for feasible route selection and resource reservation. Location-based routing, including predictive QoS routing, is now an active area of research. Clearly, QoS is a realistic goal to pursue for static ad hoc networks, e.g., a classroom setting. It is equally clear that considerable additional work is necessary to understand better both the specific conditions and extents under which various QoS objectives could be satisfied for dynamic ad hoc networks in the real world. Indeed, guaranteeing QoS in such a network may be impossible if the nodes are too mobile. Even the size of the ad hoc network becomes an issue beyond a certain level, because of the increased computational load and difficulties in propagating network updates within the given time bounds. Minimization of power consumption and QoS support do not appear to be mutually consistent objectives at the current "state of the art." Will the network have to be treated, as some have already suggested [16], as some form of a hierarchically ordered collection of subnetworks where, at each level, the pertinent size is not an issue? Is such an ordering always possible? The challenges increase even more for those ad hoc networks that, like their conventional wireless counterparts, support both best-effort services and those with QoS guarantees, allow different classes of service, and are required to inter-work with other wireless and wire-line networks, both connection oriented and connectionless. Algorithms, policies, and protocols for coordinated admission control, resource reservation, and routing for QoS under such models are only beginning to receive attention. In the latter context, the use of pre-emptive priority of control messages, class of service mechanisms, and segregation of dedicated resources for QoS-robust ad hoc networking offers promising areas of investigation. The general issue of QoS-robustness is yet uncharted territory. The same is also true for accommodating traffic with multiple priorities, including pre-emptive priorities.

Secure QoS routing introduces an entirely new dimension to the existing challenges; I have not addressed this issue. It is obvious that any sufficiently sustained denial-of-service attack, depending on the amount of resources it will cause the network to waste, will destroy QoS routing, and even worse, cause an ad hoc network to become combinatorially unstable. While secure routing is now an active area of research, QoS issues are yet to be addressed explicitly in this connection. Little has appeared so far [3] on the preservation of QoS guarantees under various failure conditions in ad hoc networks as a specific area of study. The development of suitable overload handling policies, algorithms, and protocols for preserving QoS in a mobile ad hoc network is also an open area. Performance and scalability studies of ad hoc networks with QoS constraints remain an open area of research, although important results are now appearing in this area for general routing issues associated with such networks. The indispensable issues of performance and scalability remain open for all secure routing protocols. Comprehensive performance studies are a pre-requisite to making secure QoS useful for mobile ad hoc networks. Since none exists at all, the goals for simultaneously meeting security and QoS objectives for mobile ad hoc networks offer exceptionally challenging research opportunities.

Support of multicast services, such as video conferencing, is one of the principal attractions for ad hoc networks. I have not even mentioned the manifold complex issues of adding QoS support to multicasting in mobile ad hoc networks.

Much work remains to be done on cost-effective implementation issues to bring the promise of ad hoc networks within the reach of the public. References [18 and 19] have some additional information on this topic.

6.6 References

[1] K. Wu and J. Harms, "QoS support in mobile ad hoc networks," *Crossing Boundaries*, vol. 1, no. 1, Fall 2001, pp. 92–106.

[2] S. Chen and K. Nahrstedt, "An overview of quality-of-service routing for the next generation high-speed networks: problems and solutions," *IEEE Network*, Nov.–Dec. 1998, pp. 64–79.

[3] S. Chen, "*Routing Support for Providing Guaranteed End-to-End Quality-of-Service*," Ph.D. thesis, Urbana, Illinois: University of Illinois at Urbana-Champaign, 1999.

[4] For extensive additional information on routing in mobile ad hoc networks, including QoS routing, see, inter alia, the web pages of the MONET group (including K. Nahrstedt: http://cairo.cs.uiuc.edu/~klara/home.htm, Shigang Chen: www.cise.ufl.edu/~sgchen, Baochun Li, now at U. of Toronto: www.eecg.toronto.edu/~bli/research.html), the CEDAR group (including V. Bharghavan: http://timely.crhc.uiuc.edu/index.html) of University of Illinois at Urbana-Champaign, of the Terminodes project at the Swiss Federal Institute of Technology (including: J.-P. Hubaux: http://people.epfl.ch/jean-pierre.hubaux), and the INSIGNIA team at Columbia University (including A. T. Campbell: www.cs.dartmouth.edu/~campbell/). This is only a representative list; there are many others.

[5] T.-W. Chen, "*Efficient Routing and Quality of Service Support for Ad Hoc Wireless Networks*," Ph.D. dissertation, Los Angeles, CA: University of California at LA, 1998.

[6] M. Mauve, J. Widner, and H. Hartenstein, "A survey on position-based routing in mobile ad-hoc networks," *IEEE Network*, vol. 16, Nov.–Dec. 2001, pp. 30–39.

[7] E. M. Royer, and C.-K. Toh, "A review of current routing protocols for ad hoc mobile wireless networks," *IEEE Personal Commun.*, Apr. 1999, pp. 46–55.

[8] J. Broch, D. A. Maltz, D. B. Johnson, Y.-C. Hu, and J. Jetcheva, "A performance comparison of multi-hop wireless ad hoc network routing protocols," *Proc. 4th Annual ACM/IEEE Intl. Conf. on Mobile Computing and Networking (MobiCom '98)*, Dallas, TX, Oct. 25–30, 1998, pp. 85–97.

[9] J.-P. Hubaux, L. Buttyán, and S. Čapkun, "The quest for security in mobile ad hoc networks," *Proc. 2001 ACM Intl. Symp. on Mobile Ad Hoc Networking and Computing (MobiHoc '01)*, 2001, pp. 146–155.

[10] S. Chakrabarti and A. Mishra, "Quality of service in mobile ad hoc networks," in *CRC Handbook of Wireless Ad Hoc Networks*, (M. Ilyas, Editor), CRC Press, 2002.

[11] C. E. Perkins and P. Bhagwat, "Highly dynamic destination-sequenced distance vector routing," *ACM SIGCOMM Computer Commun. Rev.*, Oct. 1994, pp. 234–244.

[12] D. B. Johnson and D. A. Maltz, "Dynamic source routing in ad hoc wireless networks," in *Mobile Computing* (T. Imielinski and H. Korth, Editors), Kluwer Academic Publishers, 1996, pp. 153–181.

[13] C. E. Perkins and E. M. Royer, "Ad-hoc on-demand distance vector routing," *Proc. 2nd IEEE Workshop on Mobile Computing Systems and Applications (WMCSA '99)*, Feb. 1999, pp. 90–100.

[14] Z. J. Haas and M. R. Pearlman, "Determining the optimal configuration for the zone routing protocol," *IEEE J. Select. Areas Commun.*, vol. 17, no. 8, Aug. 1999, pp. 1395–1414.

[15] E. Crawley, R. Nair, B. Rajagopalan, and H. Sandick, "A Framework for QoS-based Routing in the Internet," RFC 2386, http://www.ietf.org/rfc/rfc2386.txt, Aug. 1998.

[16] R. Ramanathan and M. Steenstrup, "Hierarchically-organized, multihop mobile wireless networks for quality-of-service support," *ACM/Baltzer Mobile Networks and Applications*, vol. 3, no. 1, Jun. 1998, pp. 101–119.

[17] S. H. Shah and K. Nahrstedt, *Predictive Location-Based QoS Routing in Mobile Ad Hoc Networks*, Technical Report UIUCDCS-R-2001-2242 – UILU-ENG-2001-1749, Dept. of Computer Science, University of Illinois at Urbana-Champaign, September, 2001; in *Proc. IEEE Intl. Conf. Commun. (ICC '02)*, New York, NY, Apr. 28-May 2, 2002.

[18] S. Chakrabarti and A. Mishra, "QoS issues in ad hoc wireless networking," *IEEE Communication Magazine*, vol. 39, no. 2, Feb. 2001, pp. 142–148.

[19] S. Chakrabarti and A. Mishra, "Quality of service challenges for wireless mobile ad hoc networks," *Wireless Communication and Mobile Computing*, vol. 3, no. 4, Sep. 2003, pp. 921–945.

7

Secure routing

In an ad hoc network, each node is expected to forward the packets of its immediate neighbor to a node closer to destination. Without cooperation of the nodes in the neighborhood, a packet cannot make its journey from a source to destination. If the neighboring nodes are selfish or compromised, then the correct forwarding of the packets through them may not be possible. Compromised nodes often subvert the underlying routing protocol in such a way that a packet gets forwarded to an arbitrary destination, where packets may be subjected to content modification, identity tampering, or simply dropped. This chapter examines the problem of securing the routing protocols of ad hoc networks.

7.1 Security aware routing

The desirable properties of a secure route, which are timeliness, ordering, authentication, authorization, data integrity, confidentiality, and non-repudiation are summarized in Table 7.1. The table also indicates the well known techniques that are often employed in practice in achieving these properties in a routing protocol. For example, time stamps are used to ensure timeliness and sequence numbers are used in packet headers to ensure ordering of the routing messages.

The route discovery process is an integral part of a routing protocol, which finds paths between a source–destination pair. When a route discovery process is initiated to find a path that satisfies certain specific criteria such as QoS constraints and if such a route is indeed found, then such a routing protocol is known as a QoS-aware routing protocol [1]. Similar ideas have been developed in the context of security. For example, a routing protocol that discovers a path satisfying a minimum set of security constraints can be called a security-aware routing protocol. Indeed, such a security-aware routing (SAR) protocol has been proposed for ad hoc networks in [2]; this computes secure routes

Table 7.1 *Security aware ad hoc routing properties*

Property	Techniques
Timeliness	Time stamp
Ordering	Sequence number
Authenticity	Password, certificate
Authorization	Credential
Integrity	Digest, digital signature
Confidentiality	Encryption
Non-repudiation	Chaining of digital signatures

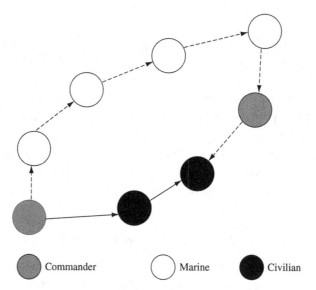

Figure 7.1 Security-aware routing

meeting some quantitative goals. The motivation for the SAR protocol stems from the recognition of facts that in certain applications, such as military, finding a route with specific security attributes or trust levels is more relevant than finding the shortest route between the two end points. An example of such a network is shown in Fig. 7.1.

Figure 7.1 illustrates a battlefield scenario in which civilians, marines, and the commanders are participating. In this network, two commanders have a secure path to themselves. Over some period of time, commanders perceive that some civilians that are part of the mission are defected or have become compromised and, as a result, decide to discover another route that consists of soldiers only. It turns out that the route through the soldiers may be more secure but it may not be the shortest path. A discovered path that is passing through the soldiers is called a security-aware route. Using security-aware routing, the commanders can find alternate routes around the compromised nodes that are more secure. By modifying the route discovery procedure of an ad hoc routing protocol, such paths could be discovered. If the route discovery fails to find such a path, then either security requirements need to be relaxed so that a new path search could be initiated or a decision is made to postpone the planned communication for a certain period of time. The following paragraphs take a closer look at SAR, albeit briefly.

7.1.1 Security-aware routing

The SAR protocol uses an AODV routing protocol for experimentation with secure routing. But the ideas are equally applicable to any other on demand ad hoc routing protocol such as DSR. Since this book has not described any routing protocol, a brief operational description of AODV is justified, and is given below. Reference [1] is an excellent source for ad hoc routing protocols.

In AODV, when a node intends to communicate with a destination node, it broadcasts a route-request message (RREQ) to its neighbors, and its neighbors propagate the message to their neighbors: as a result the RREQ ultimately reaches the destination. While moving closer to the destination, if the RREQ message finds a node that has a path to the destination, then this node creates a route-reply message (RREP) and sends it to the source node by using the path that the RREQ message used. This forwarding process is called reverse path forwarding. The RREQ message creates this path by inserting the identities of all the nodes that it encounters while traversing towards the destination.

In the security-aware routing protocol, the security metric is embedded in the RREQ packet. Upon receiving an RREQ packet, the node verifies whether it has the ability to provide the required security. If it does, the packet is forwarded to the next hop, otherwise the RREQ packet is dropped. Upon finding a path that has a desired security, the destination node or any other intermediate node creates an RREP packet and sends it to the source.

The discovered paths in SAR are differentiated in terms of the quality of security metrics. Two quality of security metrics are identified in SAR that deal with (a) the trust levels or hierarchy and (b) the security capability of a node. These metrics are discussed below.

(1) *Trust hierarchy* Security-aware routing supports a hierarchy of trust levels among the routes that are available using the route discovery process. In a hierarchical system, data packets of an equal or higher trust level can flow through a path of a certain trust level.

(2) *Security capabilities* The second quality of security metric deals with the abilities of a node in handling security-related functions such as encryption, decryption, verification of digital signatures, etc. The route discovery process in SAR finds paths through nodes that have desired capabilities necessary for a specific trust level. In SAR, the trust level and security capabilities of a node are not subject to change. Because of this constraint, a node is not able to change its own trust level or that of an RREQ message that it forwards.

7.1.2 *Operation of security-aware routing*

Security-aware routing advocates modifying the AODV protocol, particularly the RREQ and RREP messages, while preserving most of the essential characteristics of AODV. The AODV protocol incorporating these changes has been called an SAODV (security-aware AODV protocol). The proposed modifications to RREQ and RREP are discussed below.

(1) *SAR-RREQ* SAODV adds three new fields to the original AODV to make it secure. These are:
   (i) The first field, *RQ_SEC_REQUIREMENT*, specifies the security required by the sender for the route that he or she is seeking. The field can be simple integer values reflecting the needs of the application. For example, the integers 1, 2, and 3 can denote different trust levels.
   (ii) The second new field is called *REQ_SEC_QOP_VECTOR* and contains the security capabilities that are required of nodes in the discovered path.
   (iii) The third field, *RQ_SEC_GUARANTEE*, deals with the minimum security discovered by the RREQ among all paths. This field can also be represented using an integer and will represent the minimum of the security levels of the participating nodes. This information is copied into RREP by the destination and returned to the sender.
(2) *SAR-RREP* There is only one change proposed to RREP in SAR and that is copying the *RQ_SEC_GUARANTEE* field of RREQ to RREP, which allows the sender to know the available security over the entire path. Security-aware routing also recommends copying this field in the routing tables of the nodes on the path.

7.1.3 *Route discovery in SAODV*

With the proposed changes to RREP and RREQ the route discovery operation gets slightly modified over a conventional AODV. The modified route discovery process works as follows:

(1) The source node sets the RQ_SEC_REQUIREMENT field to the desired level. It also sets the appropriate bits in the RQ_SEC_QOP_VECTOR, which defines the expected security capabilities of the nodes in the route.

(2) The sender broadcasts the RREQ packet.

(3) Upon receiving the broadcast packet, the node checks to see whether it can satisfy the security requirements stipulated in the packet.

(4) If it can, then it performs the first test on the security level and the second test on the QoP bit.

(5) The RREQ packet is forwarded to the neighbors and fields are updated.

(6) If an intermediate node cannot satisfy the security requirements, the packet is dropped.

(7) In the event that some nodes get captured or compromised, SAODV tries to get the cooperation of nodes by encrypting the RREQ headers or by adding digital signatures and distributing keys to nodes that are at the same trust level. If a node with a set of keys and credentials is compromised, the adversary can assume the identity of the compromised node but cannot do more harm than the compromised node was originally allowed to do.

(8) If the RREQ message arrives at the destination, it signals the presence of a desired path between a sender and the intended receiver. The destination sends an RREP back to the sender including the information about the maximum security that could be achieved over this route.

The reference [2] includes performance evaluation results for SAR under different security scenarios. Interested readers should consult it for additional details.

7.2 Secure distance-vector routing protocols

The ad hoc on-demand distance-vector (AODV) routing protocol is a unicast, reactive routing protocol for mobile nodes in ad hoc networks [3]. It enables multi-hop routing and the nodes in the network maintain the topology dynamically only when there is traffic. The basic AODV does not specify any security mechanisms in its operation as a result; it is insecure. An approach to secure distance vector routing protocol for the Internet, which is a distributed, asynchronous routing protocol, is described in [4]. While the ideas presented in the paper are related to the Internet, they are equally applicable to the ad hoc routing protocols if they use distance vectors, such as AODV. The following paragraphs present the key ideas from the paper.

An intruder, in general, has the capability to fabricate, replay, monitor, modify, or delete any of the traffic flowing in its vicinity. From the routing perspective, the following types of intruders are of significance.

Masquerading A node is called masquerading when an intruder forges the identity of a legitimate node. Internet-protocol spoofing is one of the means of acquiring an authorized identity.

Subverted A node becomes subverted when a legitimate node is forced to violate the routing protocol in operation.

Unauthorized A node becomes unauthorized when a node that has not been assigned the role of a router acts as a router and participates in the route computation process.

Subverted links This scenario occurs when an intruder gains access to the wireless link and then manipulates the information flowing through it.

A strategically placed intruder in any of the roles discussed above can create several vulnerabilities for the network, such as compromising the network, disrupting the routing, creating denial-of-service attacks and subverting the routing messages that are exchanged by the nodes, etc. In a routing protocol, there are two types of message that are exchanged among the nodes including the source and the destination.

(1) Messages that are forwarded to the neighboring nodes; these are routing messages;
(2) Messages that are routing updates and need to be forwarded to other remote nodes, including the neighbors.

The countermeasures that are proposed for these two types of communication to make them secure require certain modifications to the routing protocol messages, as shown in Fig. 7.2. The suggested changes are the inclusion of

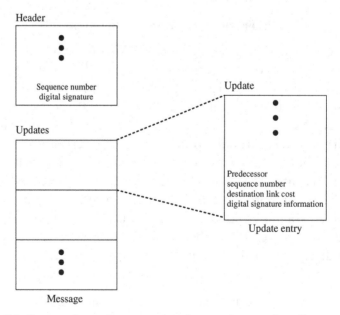

Figure 7.2 Proposed routing message changes in securing distance-vector protocols

the sequence number and the digital signature in the header part of a routing message. Similarly, changes to update messages consist of inclusion of predecessor node identity, sequence number, and the digital signature.

This routing protocol performs route computation on a per-destination basis, and it maintains information about the second-to-last network with distance information from each neighbor to every destination in the network. To ensure the authenticity and integrity of the information, the originating note digitally signs the unchanging fields of each update it generates. An IP address of the originating node is added to each update to allow receiving nodes to validate the signature. The update provides protection against the compromised nodes that have the cryptographic keys.

In the following, the proposed countermeasures are briefly described.

(1) *Routing message protection* The routing message digital signature and sequence numbers provide authentication and integrity services to routing messages, which is the first type of communication. As a rule, a sequence number is included in each routing message. This number is initialized to zero in the beginning and it is incremented with each message that is sent by the source node. When a message is skipped or a duplicate is received, as detected by the sequence number, the session is reset. The sender signs each message using its digital signature, which provides authentication and the message integrity to some extent. The message is dropped if it is corrupted.

(2) *Routing update protection* The routing updates are sent by a node to other remote nodes when the originating node sends such an update for a specific destination. The three countermeasures proposed for protecting routing updates are:

   (i) *Add sequence number to updates* This is needed to protect against the replay of old routing information. The new sequence information is generated for each route but the updates to this route have the same sequence number or time stamp for all of them. The updates received by a remote node for a given destination are considered valid if their sequence information is greater than or equal to the current sequence information. An invalid update is dropped.

   (ii) *Add predecessor information to updates* A typical distance-vector routing update message consists of one or multiple entries each specifying a destination and a distance to the destination. A node receiving the update message has to verify the validity and the authenticity of distance to a destination. By including the predecessor information in the update message, it is possible to verify the validity and the integrity of the entire path iteratively by going back to the predecessor of each node on the entire path.

   (iii) *Digitally sign updates* To provide authenticity and integrity to the routing information, the originating node digitally signs the fields which do not change, e.g. destination, predecessor, and sequence information fields. The distance field of the update changes by hop, so it is not signed by the node. To allow the receiving nodes to update the signature, an IP address of the originating node is added to each update. The signatures are used to validate a candidate path to a destination before that path is selected for use.

For additional details, such as evaluation of the effectiveness of the counter-measures and cost analysis, the reader should refer to [4].

7.3 Mitigating routing misbehavior

Sergio Marti *et al.* discussed two techniques that improve throughput of an ad hoc network in the presence of compromised nodes that agree to forward packets but fail to do so [5]. A node may misbehave because it is overloaded, selfish, malicious, or broken. An overloaded node lacks the CPU cycles, buffer space, or available network bandwidth to forward packets. A selfish node is unwilling to spend battery life, CPU cycles, or available network bandwidth to forward packets not of direct interest to it, even though it expects others to forward packets on its behalf. A malicious node could launch a denial of service attack by dropping packets. A broken node might have a software fault that prevents it from forwarding packets.

To mitigate the decrease in the throughput due to these nodes, the authors use a watchdog that identifies misbehaving nodes and a pathrater that helps routing protocols to avoid these nodes. When a node forwards a packet, the node's watchdog verifies that the next node in the path also forwards the packet. The watchdog does this by listening promiscuously to the next node's transmissions. If the next node does not forward the packet, then it is flagged as misbehaving by the "watchdog." The pathrater, run by each node in the network, combines knowledge of misbehaving nodes with link reliability data to pick the route most likely to be reliable. Each node maintains a rating for every other node in the network it knows about. It calculates a path metric by averaging the node ratings in the path.

The watchdog operation is illustrated in Fig. 7.3. Suppose there exists a path from node S to D through intermediate nodes A, B, and C. Node A cannot transmit all the way to node C, but it can listen in on node B's traffic. Thus, when A transmits a packet for B to forward to C, A can often tell if B transmits the packet. If encryption is not performed separately for each link, which can be expensive, then A can also tell if B has tampered with the payload or the header.

Every time a node fails to forward the packet, the watchdog increments the failure tally. If the tally exceeds a certain threshold bandwidth, it determines

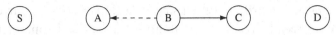

Figure 7.3 Watchdog operation

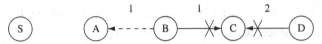

Figure 7.4 Node A does not hear B forward packet 1 to C, because B's transmission collides at A with packet 2 from source S

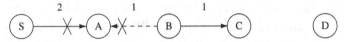

Figure 7.5 Node A believes that B has forwarded packet 1 on to C, though C never received the packet, owing to a collision with packet 2

that the node is misbehaving. The watchdog technique has advantages and weaknesses. The routing protocol with the watchdog has the advantage that it can detect misbehavior at the forwarding level and not just the link level. The watchdog's weaknesses are that it might not detect a misbehaving node in the presence of the following:

(1) Ambiguous collisions: prevents A from overhearing the transmission from B, as shown in Fig. 7.4;
(2) Receiver collisions: node A can only tell whether B has sent packet, but it cannot tell if C received it or not, as shown in Fig. 7.5;
(3) Limited transmission power: a misbehaving node could limit its transmission power such that the signal is strong enough to be overheard by the previous node but too weak to be received by the true recipient;
(4) False misbehavior: this occurs when one node falsely reports other nodes as misbehaving;
(5) Partial dropping: a node can circumvent the watchdog by dropping packets at a lower rate than the watchdog's configured minimum misbehaving threshold.

7.4 Secure packet forwarding – the currency concept

The availability of a service in an ad hoc network is strongly dependent on the packet forwarding behavior of intermediate nodes towards the destination. In other words, nodes have to cooperate with each other and refrain from being selfish. As an incentive for cooperation, Buttyan and Hubaux [6] have developed a node reward concept using a notion of virtual currency called nuggets. Please recall that the availability of a service is one of the desired attributes of secure systems.

One approach to developing cooperative behavior among the nodes is based on using the concept of currency. The idea is that the nodes that use a service should pay for it and the nodes that provide the service should be remunerated. Each node is assumed to have a specified amount of nuggets. Since currency is at the heart of any communication, nodes are motivated on the one hand to preserve currency by not sending redundant messages and on the other to help other nodes in forwarding their packets so that they earn additional nuggets.

Two approaches to packet forwarding that use nuggets are identified in [2]. These are called the packet purse model (PPM) and the packet trade model (PTM). The *packet purse model* allows the source node to pay for the packet forwarding service. The forwarding fees are distributed among the nodes to the destination as follows:

(1) When forwarding a packet, the source node puts in enough nuggets as part of the packet, so that it reaches the destination;
(2) Each forwarding node takes one or more nuggets, depending upon the service provided from the packet, to increase its own total of nuggets;
(3) If a packet runs out of nuggets prior to reaching the destination, it is discarded.

As you may have noticed, the basic problem with this model is that it might be difficult to estimate a priori the total number of nuggets needed to reach the destination. If the estimate is lower than necessary, then the packet gets dropped prior to reaching its destination and the investment gets wasted. On the other hand, if the estimate is higher than that required, then the packet makes it to the destination but loses all the unused nuggets.

The working of the PPM model is shown in Fig. 7.6. The figure shows that there are five nodes and each of them has seven nuggets, shown on their left.

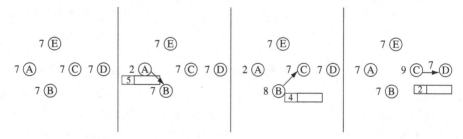

Figure 7.6 The packet purse model

(1) Let us assume that A wants to send a packet to node D.
(2) To get started, A loads five nuggets in the packet and sends the packet to B. So A has two nuggets left.
(3) B takes out one nugget as a reward for forwarding and sends the packet to C with four remaining nuggets. The number of nuggets for B increases by one to become eight.
(4) Now C forwards the packet to D by removing two nuggets: as a result the packet is sent with two nuggets. C's nuggets increase to nine.

In the *packet trade model* (PTM), the packet does not carry nuggets. The intermediate node buys the packet for a certain amount of nuggets from the previous node and sells it to the next intermediate node for more nuggets (a higher price). A node that is providing a service increases its currency as the packet moves towards the destination. Ultimately it is the destination that buys it from its predecessor node for a price and sees its account of nuggets diminished. The working of this model is illustrated in Fig. 7.7.

Let us assume that in the beginning every node had seven nuggets.

(1) Let us assume that A wants to send the packet to D, as in the previous case. A sends the packet to B for free.
(2) Upon receiving the packet, B sells the packet to C for one nugget. As a result B has one additional nugget and its total is eight, whereas C has one less nugget.
(3) Now C sells the packet to D for two nuggets. So now C has a total of eight nuggets and D has only five nuggets left.

This approach has one advantage: that the source node need not know the number of nuggets required to deliver the packet. A disadvantage of this approach is that nodes may be at liberty to flood the network. As a result some nodes may run out of currency if other nodes do not buy packets from them.

To ensure that both models work properly, certain rules need to be imposed on the nodes for each model. For example, the nodes of the packet purse model must ensure that:

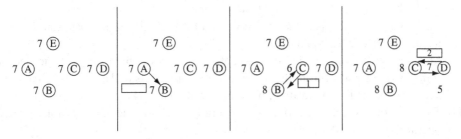

Figure 7.7 The packet trade model

(1) Nodes should not attempt nugget forgery;
(2) The originator of a packet should be denied the re-use of the nuggets that it loaded in the packet purse;
(3) A forwarding node should be prevented from taking more nuggets out of the packet than it deserves for the packet forwarding;
(4) Each intermediate node should be forced actually to forward the packet after having taken the nuggets out of it;
(5) The integrity of the packet purse should be protected during transit;
(6) The replay of a packet purse should be detected;
(7) The packet purse should not be detached from the original packet and attached to another packet and re-used. This must be impossible.

Similarly the nodes of the packet trade model should ensure that:

(1) Nodes should be prevented from re-using the nuggets already spent;
(2) A node should receive the nuggets from the next hop node only if the next hop has received the packet;
(3) An intermediate node should not be allowed to sell the same packet to other next-hop neighbors.

The solutions to the above-mentioned issues need to be efficient and computationally inexpensive to keep the overall cost of packet forwarding using the currency concept low, as compared with the overall application. The success of the currency concept relies on the availability of several features and assumptions which are listed below.

(1) Each node should have a tamper-resistant security module to secure cryptographic parameters and keys;
(2) The availability of a public key infrastructure that security modules can use for authentication and establishing secure communication links;
(3) Authentication and secure communication relies on using the public key cryptography afforded by the infrastructure;
(4) The network topology is changing slowly;
(5) Nodes are equipped with omnidirectional antennas so that each message transmitted by a node is heard by all its neighbors;
(6) The neighbor relationship is symmetric and the communication between the two neighbors is secure and reliable;
(7) Mechanisms are in place in nodes to determine the number of nuggets needed in transporting a packet from the source to a destination as well as how much a node needs to charge for forwarding, buying, or selling a packet;
(8) The network is self-organized and operates without the intervention of any operators;
(9) The nodes are greedy.

The applicability of the currency-concept-based scheme, which targets wide-area MANET, is limited by the assumption of an on-line certification authority in the MANET context. Also, the practicality of the scheme is limited by its assumptions, e.g., the computational overhead for hop-by-hop public key cryptography for each transmitted packet can be very high, and the implementation of physically tamper-resistant modules is not trivial.

7.5 Secure route discovery (SRP) and secure message transmission (SMT) protocols

As discussed earlier, the transfer of data between a source and a destination in an ad hoc network has been divided into two parts: (1) the route discovery, (2) the data transmission over the discovered paths. Most ad hoc routing protocols follow this model. From the security perspective, both the phases require security enhancements to the basic routing protocol, which has almost no security features, as discussed earlier. The first phase is vulnerable to attacks that involve impersonation of the destination, relay of outdated information, or malicious routing information, to name a few. Similarly, the second phase is exposed to tampering, packet dropping, and misdirecting the packets, etc., types of attacks. Secure routing protocol (SRP) [7] and secure message transmission (SMT) [8] are the two protocols developed to secure route discovery and the data transmission, respectively. Secure message transmission requires an SRP-like protocol for route discovery, so in the following I shall discuss SRP first, followed by SMT, albeit briefly.

7.5.1 Secure routing protocol – SRP

The SRP, in essence, provides a secure route discovery process even in the presence of malicious nodes, which have the ability to disrupt the route discovery process. For its working, the SRP assumes that the source and the destination have a shared secret key, which they use to create a secure association between themselves. In SRP, intermediate nodes do not require cryptographic validation of data or control traffic passing through them. The protocol also assumes that the secure association (SA) between the source and the destination is bidirectional so that incoming and outgoing traffic could be authenticated; besides these two nodes could authenticate themselves. The SRP also assumes that nodes are not capable of colluding within one step of the protocol execution.

In the SRP the source node (S) initiates the route discovery process with a slightly modified route request packet. The packet makes use of two identifiers, namely a query sequence number and a random query identifier. The source and the destination addresses, and the unique query identifiers, are the input along with the shared secret key to the calculation of the message authentication code (MAC) [9]. The route request packet also accumulates the identities (IP addresses) of the traversed intermediate nodes.

Intermediate nodes relay route requests, making it possible that at least one or more of these requests reach the destination. Intermediate nodes also keep track of relayed queries, so that a repeat query does not get forwarded; it is simply dropped. The destination, upon receipt of a route request, creates a route reply message. The destination calculates the MAC over the contents of the route reply message and sends the message back to the source by following the reverse path. The destination is permitted to respond to more than one packet of the same route request query.

The route discovery procedure in SRP

The source node maintains a variable called a query sequence number Q_s [10] for each destination with which it has succeeded in communicating. The sequence number allows the destination to distinguish between valid and outdated route requests. The protocol proposes to employ another variable called a query identifier Q_i, which is primarily used by the intermediate nodes. The query identifier helps in preventing the dropping of legitimate queries due to the broadcast malicious queries, as it is computationally expensive to generate Q_i. The Q_s and Q_i are both placed in the SRP header along with the MAC and other fields that the routing protocol uses. Fields in the packet header that get modified when the packet traverses towards the destination are excluded from the subsequent MAC calculations. The intermediate nodes look at the last entry (the IP address of its predecessor) in the accumulated route to the source. In a legitimate query, the last entry will belong to the neighboring node that forwarded the packet. If this node discovers that the previous node violated any of the policies or is not the correct node, the query is dropped. Otherwise the query identifier and the source and destination addresses are placed in the query table so that the node can discard the duplicate queries. New route requests are rebroadcasted and intermediate nodes insert their addresses in the route discovery packet.

The destination examines the integrity and the freshness of the queries from the nodes it had associations with, and generates route replies for all its neighbors. The reverse path serves as the source route of the reply packet. The SRP provides evidence to the source that the request had reached the destination and the reply had indeed followed the reversed path that was discovered. While the reply is traversing the reversed route, each intermediate node checks the identity of the source node, which needs to be one of its downstream nodes. If this is not the case, the packet is dropped. Upon receiving the route reply, the source checks that the route reply is indeed sent by it and examines the MAC again.

Other details of this protocol, such as route maintenance and priority based query handing, are available in [10].

7.5.2 Secure message transmission protocol

After the discovery of routes by the SRP, it is the responsibility of the SMT protocol to transmit the packet over the discovered routes, which may contain malicious nodes. Secure message transmission [8] has the ability to tolerate the presence of malicious nodes on the route. It has four key elements:

(1) End-to-end robust feedback mechanism;
(2) Dispersion of the transmitted data;
(3) Simultaneous use of multiple paths;
(4) Network adaptation.

These four elements are discussed below. Like SRP, SMT has a security association only between the source and the destination and, also like SRP, SMT does not require any specific cryptographic operations to be performed on the data in transit at the intermediate nodes. Secure message transmission advocates use of multiple paths consisting of different nodes, i.e., an intermediate node can only be on one path between a source and a destination. An example of multiple paths is shown in Fig. 7.8. In this network there are three node-disjoint paths.

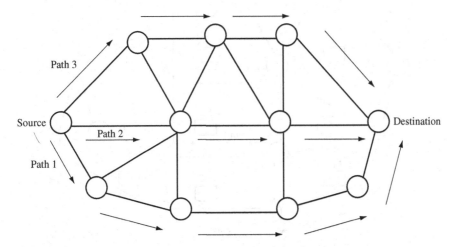

Figure 7.8 Multiple paths between source and destination

The multiple paths between the two end points are known as active path sets (APSs). The active path sets are discovered by the secure routing protocol (SRP).

A key feature of SMT is that it segments a message to be transmitted into several pieces and each piece is transmitted through each of the paths. The idea behind segmenting the message and transmitting it over multiple routes is related to the algorithm given in [11], which provides security, load balancing, and fault tolerance to data that are being forwarded. This algorithm demonstrates that it is possible to reconstruct the original message at the destination even if some pieces of the message are missing or corrupted. In other words, the message corruption and the dropping of the packets due to the presence of malicious nodes on some of the paths or unavailability of one of the paths due to breakage will have no affect on the data transmission, provided most of the pieces are received by the receiver. The segmentation and transmission of a message is shown in Fig. 7.9. Here a message has been segmented into three pieces: M1, M2, and M3. These three pieces are sent along four different paths.

Each message piece that is dispersed across the route carries a message authentication code (MAC) [9] so that the destination can verify its integrity and the authenticity of its origin. The destination validates the incoming pieces and acknowledges the successfully received pieces using acknowledgements. In SMT the process of generation of acknowledgements and their transmission is

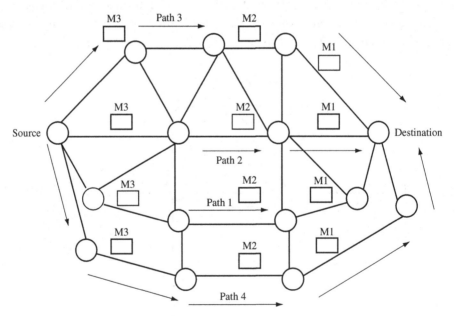

Figure 7.9 Four paths between source and destination, carrying message M

also secure and fault tolerant, which allows the source to get the authentic feedback from the destination.

In SMT, the source is authorized to update the ratings of the paths. For each successful or failed transmission, the rating of the corresponding path is incremented or decremented. Feedback of the destination is taken into account while adapting the path ratings. A path is discarded if it has failed or broken.

If a sufficient number of pieces are received at the destination, the destination starts reconstructing the message. If the message cannot be reconstructed, because of missing pieces, the destination requests the source for retransmission. Additional details about this protocol are available in [10] and [8].

7.6 Summary of security features in routing protocols and further reading

I have discussed the security features and lack thereof in the routing protocols for wireless ad hoc networks. The routing protocols must be secured from the viewpoint of authentication, integrity, non-repudiation, and privacy. These requirements can at least be partially met, for instance, by using strong encryption mechanisms, digital signatures, nonces, and time stamps. Moreover, the protection means can be optimized by analyzing potential redundancies in the routing protocol and applying efficient mechanisms, such as secret-key cryptography, hashing functions, and MACs. The use of any secret key method, however, requires a distributed, robust and secure key management service so that the necessary keys can be generated, distributed, and applied securely.

This chapter dealt with the problem of securing the route discovery and data transmission phases of a routing protocol. Some of the well known secure routing algorithms and the mechanisms they employed to provide security were discussed. However, I could not include all the algorithms that have been published. For additional reading, the reader can reference [12], which proposes a secure version of the ad hoc on demand distance vector (AODV) routing protocol using public key mechanisms to authenticate the intermediate and the destination node, and a hash chain to prevent adversaries from decreasing the hop count of the routes. Reference [13] proposed a protocol that secures a dynamic source routing protocol. The security DSR was developed as a result of the deployment of symmetric-key primitives and time synchronization to authenticate the nodes of the discovered routes. The use of public key cryptography is advocated in [14], for an AODV-like protocol.

7.7 References

[1] S. Chakrabarti and A. Mishra, "Quality of service in mobile ad hoc networks," in *The Handbook of Ad hoc Wireless Networks*, (M. Ilyas, Editor), CRC Press, 2003.

[2] S. Yi, P. Naldurg, and R. Kravets, *Security-Aware Ad-Hoc Routing for Wireless Networks*, UIUCDCS-R-2001-2241 Technical Report, Aug. 2001.

[3] C. Perkins, E. M. Royer, and S. R. Das, *Ad Hoc On-Demand Distance Vector (AODV) Routing*. IETF draft, www3.tools.ietf.org/html/draft-ietf-manet-aodv-06, 2000.

[4] B. R. Smith, S. Murthy, and J. J. Garcia-Luna-Aceves, *Securing Distance-Vector Routing Protocols*, www.isoc.org/isoc/conferences/ndss/97/smith_sl.pdf, 1997.

[5] S. Marti, T. J. Giuli, K. Lai, and M. Baker. "Mitigating routing misbehavior in mobile ad hoc networks," *6th International Conference on Mobile Computing and Networking (MOBICOM'00)*, Aug 2000, pp. 255–265.

[6] L. Buttyan and J. P. Hubaux, "Enforcing service availability in mobile ad hoc WANs," *1st MobiHoc*, Boston, Massachusetts, Aug. 2000.

[7] P. Papadimitratos and Z. J. Haas, "Secure routing for mobile ad hoc networks," *SCS Communication Networks and Distributed Systems Modeling and Simulation Conference (CNDS 2002)*, San Antonio, TX, Jan. 27–31, 2002.

[8] P. Papadimitratos and Z. J. Haas, "Secure Message Transmission in Mobile Ad Hoc Networks," in *Handbook of Wireless Ad Hoc Networks*, (M. Ilyas, Editor), CRC Press, 2003.

[9] H. Krawczyk, M. Bellare, and R. Canetti, *HMAC: Keyed-Hashing for Message Authentication*, www.rfc-ref.org/RFC-TEXTS/2104/, Feb. 1997.

[10] P. Papadimitritos and Haas, Z. "Securing mobile ad hoc networks," in *The Handbook of Ad hoc Wireless Networks*, (M. Ilyas, Editor), CRC Press, 2003.

[11] M. O. Rabin, "Efficient dispersal of information for security, load balancing, and fault tolerance," *J. ACM* vol. 36, no. 2, Dec. 2002, pp. 335–348.

[12] M. G. Zapata and N. Asokan, "Securing ad hoc routing protocol," *Proc. ACM WiSe*, Atlanta, GA, Sep. 2002, pp. 1–10.

[13] Y. Hu, A. Perrig, and D. B. Johnson, "Ariadne: a secure on-demand routing protocol for ad hoc networks," *Proc. 8th ACM Mobicom*, Atlanta, GA, Sep. 2002, pp. 12–23.

[14] K. Sanzgiri, B. Dahill, B. N. Levine, C. Shields, and E. M. Belding-Royer, "A secure routing protocol for ad hoc networks," *Proc. ICNP*, Nov. 2002, pp. 78–87.

8

Security in WiMax networks

The IEEE has created a new standard, called IEEE 802.16, that deals with providing broadband wireless access to residential and business customers, and is popularly known as *WiMax* [1]. The Worldwide Interoperability for Microwave Access (*WiMax*) is a non-profit industry trade organization that is overseeing the implementation of this standard, which is expected to replace services like Cable, DSL, and T1 line for last-mile broadband network access. It can replace these services because it has a target transmission rate that can exceed 100 Mbps. The transmission range for the WiMax devices is stated to be up to 31 miles, which also far exceeds *WiFi's* transmission range of approximately 100 meters [2,3]. With such a large transmission range, a single base station is capable of providing broadband connections to even an entire city. This chapter, briefly introduces the *WiMax* standard and then discusses the security and privacy features of such networks.

8.1 Introduction

The WiMax standard was designed with the ability to provide quality of service (QoS); as a result it can support delay-sensitive applications and services. Since it is connection oriented, it has the ability to perform per-connection QoS, allowing it to operate in both dedicated and best-effort situations.

The WiMax standard was created to meet the growing demand for broadband wireless access (BWA). This demand has proven to be challenging for service providers due to the absence of a global standard. Currently, many service providers have created proprietary solutions based on a modified version of 802.11 instead. Unfortunately these are costly solutions, which do not offer compatibility or flexibility. Some providers have tried to use 802.11 to implement a citywide deployment, despite the fact that it was designed to connect home or office computers over short distances.

147

When current WLAN technologies were examined for outdoor applications, it became clear that WiFi was not well suited for outdoor BWA applications or to provide T1 level access to businesses. A technology was needed that could operate in an outdoor environment and provide T1 level services to support data, voice, video, wireless backhaul for hotspots, and cellular tower backhaul services. The IEEE 802.16 standard was created in response to support these services, and while this standard was being defined a major emphasis was placed on the design of a physical (PHY) layer that can to support an outdoor environment, and on the media access control (MAC) layer, to provide QoS for delay sensitive applications.

8.2 Standardization and certification

A group of industry leaders (Intel, AT&T, Samsung, Motorola, Cisco, and others) have been chartered to promote the adoption of WiMax. Together they make up the WiMax Forum, which has developed a certification program for WiMax-enabled devices [2]. The goal of the forum is to define and conduct interoperability testing and award "WiMax Certified^{TM}" labels to vendor systems that pass these tests. The approach is similar to the one taken by the WiFi Alliance, which helped bring wireless LANs to the masses [1]. The WiMax certification process will also consider the European Telecommunications Standards Institute's MAN standard (HiperMAN), which will allow WiMax certified devices to work in both the US and Europe. The HiperMAN and 802.16 are both being modified in such a way that they share the same physical and medium access control layers [2].

8.2.1 Frequencies

The initial 802.16 standard specifies operation frequencies between 10 GHz and 66 GHz. The advantage of using these high frequencies is that they have more available bandwidth and less risk of interference. The disadvantage is that they require line-of-sight (LOS) environments. The 802.16a standard was adopted to provide operation in the 2 GHz to 11 GHz frequency band. The use of these lower frequencies provides the ability to support non-line-of-sight (NLOS) operation [2].

Initial WiMax deployments are expected to use the 5 GHz (license-exempt) and 2.5 GHz (licensed) frequency bands. Bands between 5.25 and 5.28 GHz will be the focus for rural areas with a low population density. For fixed wireless access, most countries have allocated the bands between 3.4 and 3.6 GHz, but the USA, Mexico, Brazil, and some Southeast Asian nations have chosen instead the bands between 2.5 and 2.7 GHz. There are also bands of interest smaller than 800 MHz, which are currently vacant or used for analog TV, due to their ability to penetrate obstacles and propagate further.

The WiMax standard will support flexible channel sizes, which will provide the ability to meet the many different channel size requirements and frequency bands from around the world. It also defines a dynamic frequency selection scheme, which helps minimize interference and increase performance.

8.2.2 Modes of operation

The WiMax standard was designed to support both point-to-point (P2P) and point-to-multipoint (PMP) topologies. While P2P can be used to support wireless network backbones, PMP is what the standard was mainly designed for. In a PMP scenario, a base station distributes traffic to many subscriber stations. To yield a high efficiency, WiMax uses a scheduling method in which base stations can only transmit in their time slots and don't contend with one another. This works quite efficiently because, unlike 802.11 hotspots, which usually have bursty traffic, stations can aggregate traffic from several computers, producing a steady flow. The WiMax standard also supports a mesh mode of operation, allowing service providers to use NLOS operation by having subscriber stations communicate directly to each other and relay traffic. Figure 8.1 illustrates the use of mesh mode in WiMax to provide a NOLS service to residential customers [2].

WiMax's design allows it to be used in many different operating environments. The ability to provide last-mile broadband access to consumers was one major consideration during development. With a focus on standardization and inoperability, WiMax may provide a low-cost solution. Figure 8.1 illustrates the possible uses of WiMax, including reliable business access, residential access, and high-speed connectivity for mobile users.

WiMax's ability to provide high transfer rates allows it to be used as a network backbone. Specifically, developers envision using WiMax as a backbone for 802.11 hotspots to provide Internet access. In this configuration users would connect to a nearby 802.11 base station. The base station would then relay the user's data to a central WiMax base station, which is connected to the Internet. This would provide citywide Internet access without having to run cables to each 802.11 hotspot.

Another access method would be to allow users to connect directly to the WiMax base station, allowing citywide Internet access with a single point of attachment, without the need for any 802.11 base stations. Although possible, it may not use bandwidth as efficiently as the previous example. This is because of the scheduling algorithm WiMax uses, which is designed for steady and

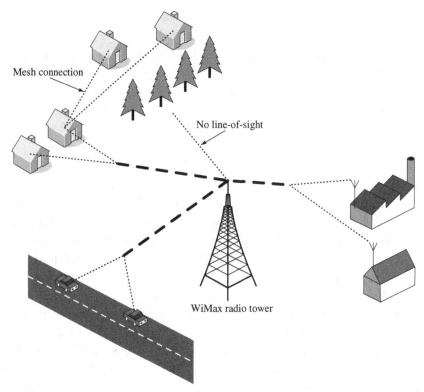

Figure 8.1 Overview of WiMax

smooth traffic and not for the bursty traffic created by individual users. Also it is likely that base stations may only have a range of 5 to 6 miles instead of 30, owing to the increased vulnerability of the links from the user mobility.

Developers also had cellular applications in mind when they designed WiMax. The first use of WiMax for cellular applications will be a tower backhaul service. Once the IEEE 802.16e standard is implemented, which is optimized to support handoffs and roaming at speeds of up to 75 mph, WiMax can be used to connect directly to cell phones and other mobile devices.

Many government agencies see the value of using WiMax for both home-land security and in emergency situations. Agencies could deploy WiMax enabled devices to monitor high valued infrastructures and transmit the infor-mation to a central operations center for processing [4]. Emergency mobile wireless networks are another important use for the government. During a disaster, where all communications have been lost, a WiMax network could be quickly set up. This would allow organizations like FEMA, Red Cross, and NATO to communicate important information that may be crucial to rescue operations [5].

8.3 Frame structure

8.3.1 The physical layer

When the 802.16 standard was introduced it had a single carrier (SC) PHY specification to support LOS operations in the 10–66 GHz frequency band. With the 802.16a amendment to the standard, changes to the PHY were needed to support the 2–11 GHz frequency band. This led to the introduction of a new single carrier PHY, a 256 point FFT OFDM PHY, and a 2048 point FFT OFDMA PHY [1]. The 802.16e amendment to the standard provides an enhanced version of OFDMA, called scalable OFDMA (SOFDMA) [2]. The SC PHY specification is designed for LOS operation in the 10–66 GHz frequency band. Both TDD and FDD configurations are supported to allow for flexible spectrum usage. The SC PHY is designed for NLOS operation in the 2–11 GHz frequency band and is based on SC technology. The OFDM PHY uses a 256 carrier OFDM and uses TDMA to provide multiple access to different subscriber stations. The OFDMA PHY uses a 2048 carrier OFDM design. Multiple access is provided by assigning a subset of the carriers to an individual subscriber station. The enhancement of the OFDMA PHY, SOFDMA, uses the values 128, 512, 1024, and 2048 to scale the number of sub-carriers in a channel [2,6].

The WiMax Forum decided that the first interoperable tests and certifications for 802.16 devices would support OFDM. While OFDMA can allocate spectrum more efficiently and reduce interference, compared with OFDM it is more complex to install and operate. Therefore, OFDMA is only required for 802.16e certified devices, where it is needed to support mobile customers. The WiMax Forum has worked with Korean standard WiBro, which uses SOFDMA, to insure that the two technologies will be interoperable. Eventually, SOFDMA will be the PHY layer of choice for indoor and mobile equipment [1,2].

8.3.2 The MAC layer

In a WiMax environment it can be difficult for subscriber stations to listen to one another. Therefore the MAC layer was designed to use a flexible frame structure in which the base station schedules subscriber station transmissions in advance. This reduces contention because subscriber stations only need to contend when they access the channel for the first time. The overall effect is increased efficiency, which allows one base station to serve a large number of subscriber stations.

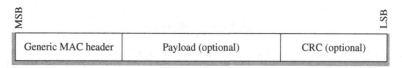

Figure 8.2 Protocol data unit (PDU) in MAC

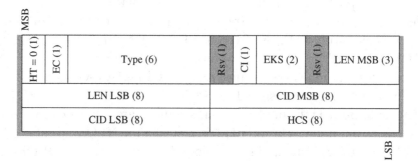

Figure 8.3 Generic MAC header

Protocol data units (PDUs) in MAC

Each protocol data unit (PDU), as seen in Fig. 8.2, comprises a generic MAC header (GMH), a payload, and an optional cyclic redundancy check (CRC). The GMH defines the contents of the payload and starts at the most significant bit (MSB). The payload consists of zero or more sub-headers and MAC service data units (SDUs). The length of the payload may vary. The CRC is optional in the SC PHY layer but mandatory for SCa, OFDM, and OFDMA PHY layers [6].

There are two formats defined for the MAC header. The GMH is used for MAC PDUs that contain MAC management messages or convergence sub-layer data. The bandwidth request header is used when requesting additional bandwidth. The two headers are distinguished by the single-bit header type (HT) field, which is zero for the generic header and one for the bandwidth request header.

The GMH, shown in Fig. 8.3, is encoded from the HT field on. The GMH is 6 bytes in length and consists of 12 fields. Two of these fields, which are 1 bit in length each, are reserved for future use. The remaining fields are defined in Table 8.1.

The type field of the GMH is used to indicate what type of sub-header or special payload is included in the message. The possible type values and corresponding meanings are defined in Table 8.2.

The bandwidth request PDU, shown in Fig. 8.4, has no payload and consists of only the header. It is 6 bytes in length and consists of 8 fields, which

Table 8.1 *Generic MAC header fields*

Name	Length (bits)	Description
Cl	1	CRC indicator 1 = CRC is included in the PDU by appending it to the payload after encryption if any 0 = No CRC is included
CID	16	Connection identifier
EC	1	Encryption control 0 = Payload is not encrypted 1 = Payload is encrypted
EKS	2	Encryption key sequence The index of the traffic encryption key (TEK) and initialization vector used to encrypt the payload. This field is only meaningful if the EC field is set to 1
HCS	8	Header check sequence An 8 bit field used to detect errors in the header
HT	1	Header type Should be set to zero
LEN	11	Length The length in bytes of the MAC PDU including the MAC header and the CRC if present
Type	6	This field indicates the sub-headers and special payload types present in the message payload

are defined in Table 8.3. Like the GMH, the bandwidth request header is encoded from the HT field on.

8.4 Point-to-multipoint (PMP) mode

In PMP mode, several subscriber stations connect to a single base station. Each subscriber station is uniquely defined by a 48 bit universal MAC address. It is used during the initial ranging process and during the authentication process so that the base station and subscriber station can verify each other's identities [6].

When a subscriber station first connects to a base station, two pairs of management connections are created between the subscriber station and the base station. An optional third pair of management connections may be created. Each pair consists of one uplink and one downlink connection, identified by a 16 bit connection ID (CID). Short, time-urgent management messages are sent over the basic connection. Longer, delay-tolerant management messages are sent over the primary management connection. Standards-based

Table 8.2 *Type encodings*

Type bit	Value
5 (MSB)	Mesh sub-header 1 = present 0 = absent
4	ARQ feedback payload 1 = present 0 = absent
3	Extended type Indicates whether the present packing or fragmentation sub-headers are extended 1 = extended 0 = not extended
2	Fragmentation sub-header 1 = present 0 = absent
1	Packing sub-header 1 = present 0 = absent
0 (LSB)	Downlink: FAST-FEEDBACK allocation sub-header Uplink: grant management sub-header 1 = present 0 = absent

HT = 1 (1)	EC = 0 (1)	Type (3)	BR MSB (11)
BR LSB (8)			CID MSB (8)
CID LSB (8)			HCS (8)

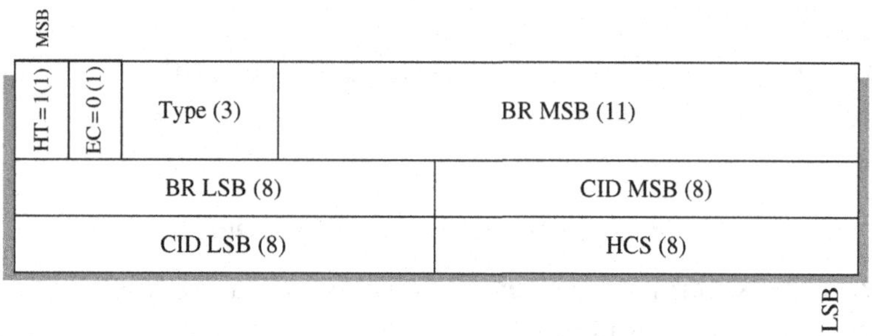

Figure 8.4 Bandwidth request header

messages (i.e., DHCP, TFTP, SNMP) are sent using the secondary management connection.

Base stations do not have to coordinate their transmissions with other stations. They simply divide time into uplink and downlink transmission periods using TDD. Downlink messages are generally broadcast. A downlink map (DL-MAP) message can be used to define access to the downlink information by defining burst start times to subscriber stations. If a DL-MAP message does not explicitly indicate a portion of the downlink for a specific

Table 8.3 *Bandwidth request header fields*

Name	Length (bits)	Description
BR	19	Bandwidth request
		The number of bytes of uplink bandwidth requested by the subscriber station. The bandwidth request is for the CID. The request shall not include any PHY overhead.
CID	16	Connection identifier
EC	1	Always set to zero
HCS	8	Header check sequence
		An 8 bit field used to detect errors in the header
HT	1	Header type = 1
Type	3	Indicates the type of bandwidth request header

subscriber station, then all subscriber stations capable of listening will listen. The subscriber stations will check the CIDs of the PDU and keep only the ones addressed to them.

Uplink transmissions to the base station are shared among subscriber stations, on a demand basis. Subscriber stations use an uplink map (UL-MAP), which is obtained from the base station, to determine when it can transmit. Four different types of uplink scheduling mechanisms are used to control contention between users and tailor the delay and bandwidth requirements of each user application. These are implemented using unsolicited bandwidth grants, polling, and contention procedures. Performance can be optimized by using different combinations of these bandwidth allocation techniques.

8.5 Mesh

In the mesh mode, subscriber stations can transmit to each other directly, allowing traffic to be routed through subscriber stations if two nodes cannot directly communicate. The advantage of the mesh mode is that it can provide NLOS communication for stations using higher frequency bands. This is accomplished by marking a node as a mesh base station if it has a direct connection to backhaul services outside the mesh network. Otherwise, it is marked as a mesh subscriber station. Traffic can then flow from mesh subscriber stations to mesh base stations, then out of the mesh network and vice versa [6].

As with the PMP mode, each node is uniquely defined by a 48 bit universal MAC address. It is used during the network entry process and during the authentication process where the entry node and the network verify each other's identities. Once a node is authorized to the network, it requests a 16 bit node identifier (node ID) from the mesh base station. This node ID is used to identify nodes during operation.

A node views other stations in its mesh network in three different ways. Neighbors are stations to which the node has a direct link, which are considered to be "one hop" away. A neighborhood consists of all the neighbors of a node. Finally, an extended neighborhood contains all the neighbors of the neighborhood in addition to the neighborhood itself.

All communications within a mesh network are in the context of a link. Eight bit link identifiers (link IDs) are used to address nodes in the local neighborhood. Each link established between a node and its neighbors is assigned a link ID. As neighboring nodes establish new links, link IDs are communicated during the link establishment process. All data transmissions between two nodes use the same link.

The mesh mode uses two types of scheduling; distributed and centralized. In distributed scheduling, all the nodes must coordinate their transmissions in their extended neighborhood. This can be accomplished by having every node broadcasting its schedule (available resources, requests, and grants) to all their neighbors. Schedules may also be established by directed uncoordinated requests and grants between two nodes. Before transmission, a node must ensure that it will not cause collisions with the transmissions scheduled by any other node in its extended neighborhood.

In the centralized scheduling, resource request from all the mesh subscriber stations within a certain hop range are gathered by the mesh base station. The base station determines the amount of resources it wishes to grant on each link in the network, and communicates the grants to all the mesh subscriber stations in the hop range.

8.6 Quality of service

The WiMax standard was designed with QoS in mind to provide low latency for delay sensitive services and data prioritization. Quality-of-service support resides within the MAC layers of the base station and the subscriber stations. The base station contains a packet queue for each downlink connection. It uses the QoS parameters and the status of the queues to determine which queue to use for the next SDUs to be sent. The subscriber station has similar queues for uplink connections [7].

Bandwidth is granted to the subscriber stations from the base stations when it is needed. Subscriber stations can request bandwidth a few different ways. Using unsolicited granting, during the setup of an uplink connection, subscriber stations request a fixed amount of bandwidth on a periodic basis. Once the connection is complete the subscriber stations cannot request any more bandwidth. The base station can use broadcast polls to determine whether subscriber stations need bandwidth. An issue arises when two or more stations respond to the same poll, causing a collision. After collision, nodes follow an exponential backoff algorithm and wait to respond again. Bandwidth requests can also be piggy-backed on a PDU sent from the subscriber station.

The base station's uplink scheduler uses the bandwidth requests to estimate the remaining backlog at each uplink connection. It uses this knowledge and the set of QoS parameters to determine future uplink grants. While the bandwidth requests are per connection, the base station grants uplink capacity to each subscriber station as a whole. Therefore, the subscriber station also implements a scheduler within its MAC to allocate its uplink bandwidth between its connections.

8.7 Security features in WiMax

WiMax security has two goals; one is to provide privacy across the wireless network and the other is to provide access control to the network. Privacy is accomplished by encrypting connections between the subscriber station and the base station. The base station protects against unauthorized access by enforcing encryption of service flows across the network. A privacy and key management (PKM) protocol is used by the base station to control the distribution of keying data to subscriber stations. This allows the subscriber and base stations to synchronize keying data. Digital-certificate-based subscriber station authentication is included in the PKM to provide access control [6].

8.7.1 Security associations

A security association (SA) is the set of security information a base station and one or more of its client subscriber stations share in order to support secure communication across a WiMax network. WiMax uses two different types of SA, data and authorization [6,8].

There are three different types of data SA: primary, static, and dynamic. Primary SAs are established by the subscriber stations during their initialization process. The base station provides the static SAs. Dynamic SAs are established and eliminated as needed for service flows. Both static and dynamic SAs can be shared among multiple subscriber stations [6].

Table 8.4 shows the contents of a data SA. The SAID is used to identify the data SA uniquely. The encryption cipher defines which method of encryption

Table 8.4 *Contents of data security associations*

Data SA
16 bit SA identifier (SAID)
Encryption cipher to protect the data exchanged over the connection
Two traffic encryption keys (TEKs): one for current operation and another for when the current key expires
Two 2 bit key identifiers, one for each TEK
TEK lifetime. The minimum value is 30 minutes and the maximum value is 7 days. The default is half a day.
Initialization vector for each TEK
Data SA type indicator (primary, static, dynamic)

will be used to encrypt data. Initially the IEEE 802.16 standard defined the use of the data encryption standard (DES) in cipher block chaining (CBC) mode. Later, in the IEEE 802.16e revision, more modes were defined. Section 3.4 of this revision covers data encryption in detail.

Traffic encryption keys (TEKs) are used to encrypt data transmissions between the base stations and subscriber stations. The data SA defines two TEKs, one for current operations and a second to be used when the current one expires. Two TEK identifiers are included, one for each key. A TEK lifetime is also included to indicate when the TEK expires. The default lifetime is half a day, but it can vary from thirty minutes to seven days.

The data encryption standard in CBC mode requires an initialization vector to operate. Therefore, one for each TEK is included in the data SA. Both initialization vectors are 64 bits in length to accommodate the 64 bit block size used in DES encryption.

The data SA type is also included to indicate whether it is a primary, static, or dynamic data SA.

Data SAs protect transport connections between one or more subscriber stations and a base station. Subscriber stations typically have one SA for their secondary management channel and either one SA for both uplink and downlink transport connections or separate SAs for uplink and downlink connections. For multicasting, each group requires an SA to be shared among its members; therefore the standard lets many connection IDs share a single SA [8].

Authorization SAs are shared between a base station and a subscriber station. They are used by the base station to configure data SAs for the subscriber station [8].

Table 8.5 shows the contents of an authorization SA. An X.509 certificate is included, which allows the base station to identify the subscriber station.

Table 8.5 *Contents of authorization SAs*

Authorization SA
X.509 certificate identifying the subscriber station
160 bit authorization key
4 bit authorization key identifier
Authorization key lifetime. The minimum value is one day and the value maximum is 70 days. The default is seven days.
Key encryption key (KEK) for distributing TEKs
Downlink hash function–base message authentication code (HMAC) key
Uplink HMAC key
List of authorized data SAs

Section 3.2.2 of IEEE 802.16e goes into detail about X.509 certificates and how they are used.

The 160 bit authorization key (AK) is included to allow the base station and subscriber station to authenticate each other during TEK exchanges. Section 3.3.2 of IEEE 802.16e describes the TEK exchange process. A 4 bit AK identifier is used to distinguish among different AKs. An AK lifetime is also included to indicate when the AK expires. The default lifetime is seven days, but it can range from one day to seventy days.

Key encryption keys (KEKs) are used to encrypt TEKs during the TEK exchange process. Two KEKs are required for the encryption process and are derived from the AK. The KEKs are computed by first concatenating the hex value 0×53 repeated 64 times and the AK. Then the SHA-1 hash of this value is computed, which outputs 160 bits. Finally the first 128 bits of the output are taken and divided into two 64 bit TEKs. These two TEKs are included in the authorization SA.

Two hashed message authentication code (HMAC) keys, one for uplink and one for downlink, are included to allow for the creation of HMACs during the TEK exchange process. The uplink key is used to create an HMAC of messages to be sent, while the downlink key is used to create an HMAC of messages received, allowing the receiver to authenticate the message. The uplink key is obtained by concatenating the hex value $0 \times 3A$ repeated 64 times and the AK, then computing the SHA-1 hash of this value, creating a 160 bit HMAC key. The downlink key is computed in the same fashion but the hex value $0 \times 5C$ is concatenated with the AK instead.

A list of authorized data SAs is also included in the authorization SA, which provides the subscriber station with knowledge of the data SAs it can request.

8.7.2 Authentication

Hashed message authentication code (HMAC)

Hashed message authentication codes are used to provide message authentication. By using HMACs, the receiver can verify who sent the message. This is possible because the sender creates an HMAC of the message it wishes to send using a key known only by the sender and receiver. When the receiver gets the message, it computes its own HMAC of the message using the same key and compares the one it computed with the one it received from the sender. If the HMACs match, then the sender is confirmed.

The HMACs are created as a function of a key and the message. Figure 8.5 illustrates the HMAC creation process. First the hash key, defined in the authorization SA, is exclusive-ored (XORed) with an ipad, which is the byte 0×36 repeated 20 times to match the size of the hash key. This 160 bit value is appended to the beginning of the message, which is then hashed. The IEEE 802.16 standard defines the use of SHA-1 to compute the hash.

The hash key is then XORed with an opad, which is the byte $0\times5C$ repeated 20 times to match the size of the hash key. This 160 bit value is appended to the beginning of the output of the previous hash. These two values are then hashed to produce the HMAC.

Certificates for X.509

Certificates for X.509 are used to allow the base station to identify subscriber stations. Table 8.6 describes the required fields, as defined by the IEEE 802.16 standard. While extension data may be included, the standard does not define any [6,8].

There are two types of certificates defined: manufacturer certificates and subscriber station certificates. A manufacturer certificate, which identifies the manufacturer of the device, can be a self-signed certificate or issued by a third party. A subscriber station certificate is typically created and signed by the manufacturer of the station. It is used to identify a subscriber station and includes the MAC address of the station in the subject field. Base stations can use the manufacturer certificate to verify the subscriber station's certificate, allowing it to determine whether the device is legitimate [8].

8.7.3 Extensible authentication protocol

The IEEE 802.16e standard introduced an alternative to the authentication scheme based on X.509 certificates. This new scheme is considered to be more flexible and is based on the extensible authentication protocol (EAP) [8].

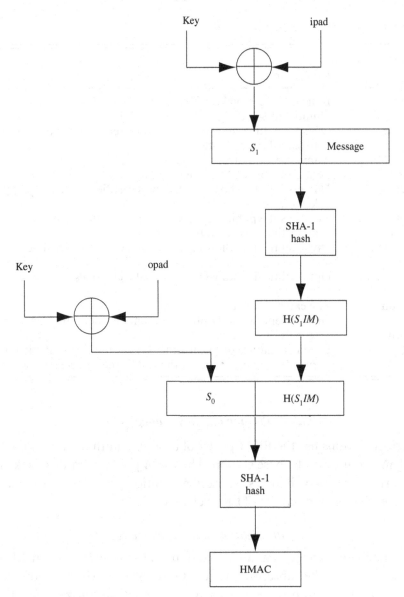

Figure 8.5 Creation of HMAC

To obtain authentication during link establishment, EAP messages are encoded directly into management frames. Two additional PKM messages, PKM EAP request and PKM EAP response, were added to transport EAP data.

Currently, EAP methods to support the security needs of wireless networks are an active area of research and, therefore, the IEEE 802.16e standard does not specify a particular EAP based authentication method to be used.

Table 8.6 *Certificate fields for X.509*

X.509 certificate fields	Description
Version	Indicates the X.509 certificate version
Serial number	Unique integer assigned by the issuing CA
Signature	Object identifier and optional parameters defining algorithm used to sign the certificate
Issuer	Name of CA that issued the certificate
Validity	Period in which certificate is valid
Subject	Name of entity whose public key is certified in the subject public key into field
Subject public key info	Contains the public key, parameters, and the identifier of the algorithm used with the key
Issuer's unique ID	Optional field to allow re-use of issuer names over time
Subject's unique ID	Optional field to allow re-use of subject names over time
Extensions	The extension data
Signature algorithm	Object identifier and optional parameters defining algorithm used to sign the certificate
Signature value	Digital signature of the abstract syntax notation 1 distinguished encoding rules, encoding the rest of the certificate

8.7.4 Privacy and key management

Subscriber stations use the PKM protocol to obtain authorization and traffic keying material form the base station. The PKM protocol can be broken into two parts. The first handles subscriber station authorization and AK exchange. The second handles TEK exchange [6].

Authorization and AK exchange

Privacy and key management authorization is used to exchange an AK from the base station to the subscriber station. Once the subscriber station receives an initial authorization, it will periodically seek reauthorization. The AK exchange is accomplished using three messages, illustrated in Fig. 8.6 [6,8].

The subscriber station initiates the exchange by sending a message containing the subscriber station manufacturer's X.509 certificate to the base station. The message is strictly informative and can be ignored by the base station. However, base stations can be configured to only allow access to devices from trusted manufacturers.

The second message is sent from the subscriber station to the base station immediately after the first message. This message is a request for an AK and a

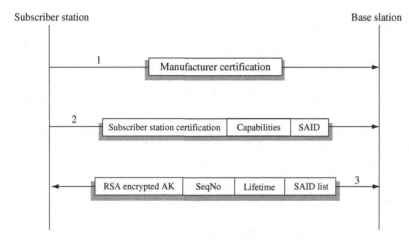

Figure 8.6 Privacy and key management authorization

list of SAIDs that identify SAs that the subscriber station is authorized to participate in. There are three parts to the message: a manufacturer-issued X.509 certificate, cryptographic algorithms supported by the subscriber station, and the SAID of its primary SA.

The base station uses the subscriber station's certification to determine if it is authorized. If it is, the base station will respond with the third message. The base station uses the subscriber station's public key, obtained from its certification, to encrypt the AK using RSA. The encrypted AK is then included in the message along with the SeqNo, which distinguishes between successive AKs, the key lifetime, and a list of SAIDs of the static SAs that the subscriber station is authorized to participate in.

Traffic encryption key exchange

Once the subscriber station has been authorized, it will establish an SA for each SAID in the list received from the base station. This is accomplished by initiating a TEK exchange. Once an SA is established, the subscriber station will periodically refresh keying material. The base station can also force rekeying if needed. Figure 8.7 illustrates the TEK exchange process [6,8].

The first message of a TEK exchange is optional and allows the base station to force rekeying. There are three parts to the message: the SeqNo refers to the AK used in creating the HMAC, the SAID refers to the SA that is being rekeyed, and the HMAC allows the subscriber station to authenticate the message.

The second message is sent by the subscriber station in response to the first message or if the subscriber station wants to refresh the keying material. There

Subscriber station   Base station

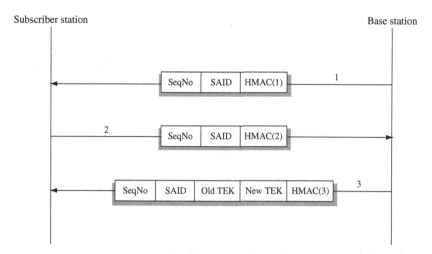

Figure 8.7 Privacy and key management traffic encryption key exchange

are three parts to the message: the SeqNo refers to the AK used in creating the HMAC, the SAID refers to either the SAID received in the first message or one of the SAs from the subscriber station's authorized SAID list, and the HMAC allows the base station to authenticate the message.

If the HMAC in the second message is valid, then the base station will send the third message. As in the first two messages a SeqNo, the SAID, and the HMAC are included. In addition to these the old TEK and a new TEK are added. The old TEK merely reiterates the active SA parameters while the new TEK is to be used when the active one expires. The base station encrypts both the old and new TEKs using triple DES in electronic code book (ECB) mode with the KEK associated with the SA.

Figure 8.8 illustrates the TEK encryption process. Section 8.7.1 described how the KEK is created. Here KEK 1 is the leftmost 64 bits of the computed KEK and KEK 2 is the rightmost 64 bits. These two keys are used in the triple DES encryption, in which the TEK is first encrypted using KEK 1. The output is then decrypted using KEK 2 and then encrypted using KEK 1. This process is performed on both the old and new TEKs to produce two encrypted TEKs.

8.7.5 Data encryption

To provide privacy for data being transmitted in WiMax networks, the IEEE 802.16 standard employed the use of DES in CBC mode. Currently DES is considered to be insecure and has been replaced by the advanced encryption standard (AES). Therefore the IEEE 802.16e standard defines the use of AES for use in encryption [8].

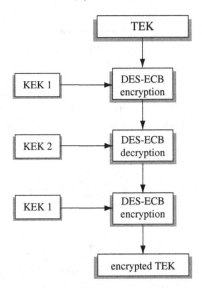

Figure 8.8 Traffic encryption key encryption process

The data encryption standard

Using the DES in CBC mode, the payload field of the MAC PDU is encrypted but the GMH and CRC are not. Figure 8.9 illustrates the encryption process.

The CBC mode requires an initialization vector (IV), which is computed by taking the XOR of the IV parameter in the SA and the content of the PHY synchronization field. The DES encryption process uses the IV and the TEK from the SA of the connection to encrypt the payload of the PDU. This ciphertext payload then replaces the original plaintext payload. The EC bit in the GMH will be set to 1 to indicate an encrypted payload and the EKS bits will be set to indicate that the TEK was used to encrypt the payload. If the CRC is included, it will be updated for the new ciphertext payload [6].

Advanced encryption standard

The IEEE 802.16e standard added the use of AES to provide stronger encryption of data. It defines the use of AES in four modes: CBC, counter encryption (CTR), CTR with CBC message authentication code (CCM), and ECB. The CTR mode is considered better than the CBC mode due to its ability to perform parallel processing of data and preprocessing of encryption blocks and the fact that it is simpler to implement. The CCM mode adds the ability to determine the authenticity of an encrypted message to CTR mode. The ECB mode is used to encrypt TEKs.

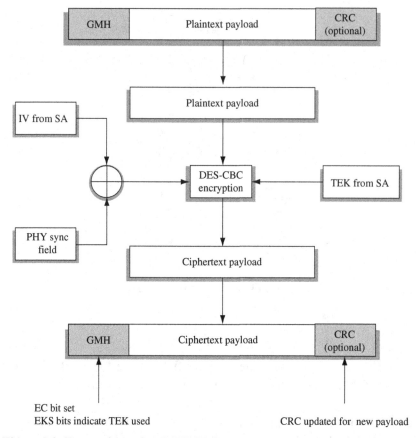

Figure 8.9 Encryption using DES-CBC

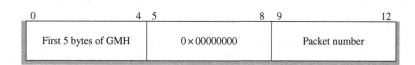

Figure 8.10 The CCM nonce

The AES in CCM mode AES-CCM requires that the transmitter construct a unique nonce, which is a per-packet encryption randomizer. The IEEE 802.16e standard defines a 13 byte nonce, as show in Fig. 8.10. Bytes 0–4 are constructed from the first five bytes of the GMH. Bytes 5–8 are reserved and are all set to 0. Bytes 9–12 are set to the packet number (PN). The PN is associated with an SA and set to 1 when the SA is established and when a new TEK is installed. Since the nonce is dependant on the GMH, modifications to the GMH can be detected by the receiver [8,9].

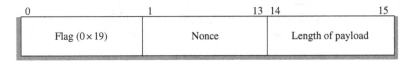

Figure 8.11 The CCM CBC block

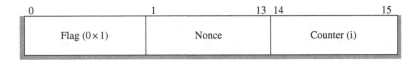

Figure 8.12 The CCM counter block

To create a message authentication code, AES-CCM uses a variation of CBC mode. Instead of using an IV, an initial CBC block is appended to the beginning of the message before it is encrypted. As seen in Fig. 8.11, the initial CBC block consists of a flag, the packet nonce, and the length of the payload.

To encrypt the payload and the message authentication code, AES-CCM uses the CTR mode. With this mode, n counter blocks are created, where n is the number of blocks needed to match the size of the message plus one block for the message authentication code (AES uses 128 bit block sizes). The first block is used for encrypting the message authentication code and the remaining blocks are used to encrypt the payload. As seen in Fig. 8.12, the counter block consists of a flag, the packet nonce, and the block number i, where i goes from 0 to n.

The message authentication code is created by encrypting the initial CBC block and plaintext payload. Figure 8.13 illustrates the message authentication code creation and subsequent encryption of the message authentication code.

The first step in creating the message authentication code is to extract the plaintext payload from the PDU and append the initial CBC block to the beginning of it. This is then encrypted using AES in CBC mode with the TEK from the SA of the connection. The last 128 bits (size of one AES block) of the encrypted output is selected to represent the message authentication code.

The sender will perform this process and then encrypt the message authentication code with the message. The receiver will decrypt the message and message authentication code and then perform the same process on the message. The receiver will then compare the message authentication code it created with the one received. If they are the same, the message is authenticated, if not the message is discarded.

Encryption of the message authentication code is accomplished by encrypting counter block 0 using AES in CTR mode with the TEK from the SA of the

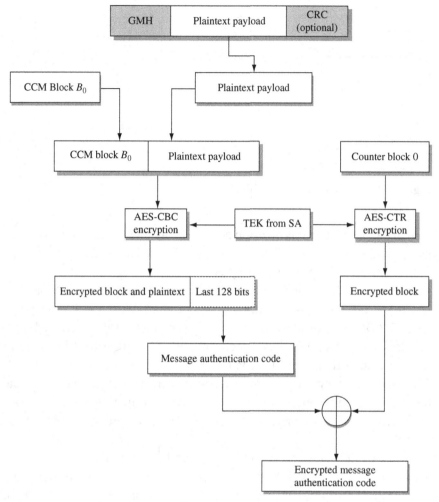

Figure 8.13 The AES-CCM message authentication creation and encryption

connection. This encrypted block is then XORed with the message authentication code to produce the encrypted version.

Payload encryption is accomplished by first encrypting counter blocks 1 through n with AES in CTR mode using the same TEK used to encrypt the message authentication code. The plaintext payload is then extracted from the PDU and XORed with the encrypted counter blocks. This produces the ciphertext payload, as show in Fig. 8.14.

The PN is then appended before the ciphertext payload and the message authentication code is appended after the ciphertext payload. This set of data then replaces the plaintext payload. The EC bit in the GMH will be set to 1 to

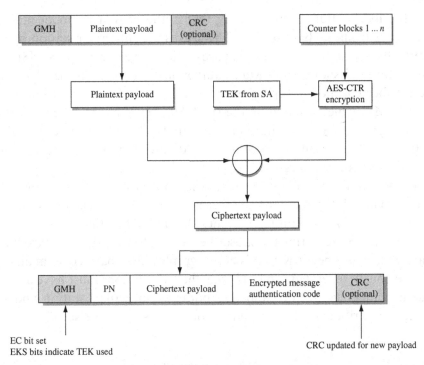

Figure 8.14 The AES-CCM payload encryption

indicate an encrypted payload and the EKS bits will be set to indicate the TEK used to encrypt the payload. If the CRC is included, it will be updated for the new payload.

8.8 Open issues

In WiMax, security threats apply to both the PHY and MAC layers. Possible PHY level attacks include jamming of a radio spectrum, causing denial of service to all stations, and flooding a station with frames to drain its battery. Currently there are no efficient techniques available to prevent PHY layer attacks. Therefore, the focus of WiMax security is completely at the MAC level [8]. In this section, I discuss some of the open security issues in WiMax networks.

8.8.1 Authorization vulnerabilities

A major vulnerability of WiMax security is the lack of a base station certificate, which is needed for mutual authentication. Without mutual authentication, the subscriber stations cannot verify that authorization protocol messages received are from the base station. This leaves the subscriber station open to forgery attacks, allowing any rogue base station to send it responses [8].

A solution to issues with WiMax's authentication and authorization procedures has been proposed. It proposes the wireless key management infrastructure (WKMI), which is based on the IEEE 802.11i standard. The WKMI is a key management hierarchy infrastructure that is based on the use of X.509 certificates, allowing subscriber stations and base stations to perform mutual authentication and key negotiation.

Authorization key generation is another concern with the authorization protocol. Though the standard assumes a random AK generation, it imposes no requirements. An additional weakness lies in the fact that the base station generates the AK, requiring the subscriber station to trust that the base station always generates a new AK that is cryptographically separated from all other AKs previously generated. To hold true, the base stations must have a perfect random number generator. Allowing both the subscriber station and base station to contribute to the AK generation could solve this issue [8].

8.8.2 Key management

A major issue with key management in WiMax is the size of its TEK identifier. Currently a 2 bit number is used, which allows only four values (0 to 3) to be represented. This causes the TEK identifier to wrap from 3 to 0 on every fourth key, leaving stations open to replay attacks, in which an attacker could re-use expired keys. To solve this issue, the TEK identifier's size needs to be increased to prevent wrapping. If the longest AK lifetime (70 days) and shortest TEK lifetime (30 minutes) is considered, then 3360 different TEKs need to be represented, which would require 12 bits be used for the TEK identifier [8]. Another issue is the TEK lifetime, which can be set anywhere between 30 minutes and 7 days, with a default of half a day. If DES in CBC mode is used for encryption with the possible lifetime values, the security of the data may be compromised. This is because DES in CBC mode becomes insecure after operating on $2^{n/2}$ blocks with the same encryption key, where n is the block size. Since DES uses a 64 bit block size, after 2^{32} blocks the encryption will be insecure. The time it takes to happen depends on the average throughput between stations. Considering the high transfer rates WiMax offers and the ability to choose a larger TEK lifetime, encryption insecurity is highly likely.

The introduction of AES in the IEEE 802.16e standard will help solve the TEK lifetime issues. Unfortunately implementation of this standard is still a way off, possibly leaving current deployments of WiMax insecure.

8.9 Summary and further reading

This chapter dealt with the introduction of a new IEEE 802.16a standard, popularly known as WiMax, as well as its security features. The chapter also included references to recent papers that have appeared in the WiMax security arena. Recently, wireless mesh networking, a most prominent application of WiMax, is attracting significant interest from academia, industry, and standards organizations. With several favourable characteristics, such as dynamic self-organization, self-configuration, self-healing, easy maintenance, high scalability, and reliable services, wireless mesh networks have been advocated as a cost-effective approach to support high-speed last mile connectivity and ubiquitous broadband access in the context of home networking, enterprise networking, or community networking. Excellent references exist on WiMax networks [10,11] that can be used to get additional information on the technology, protocols, and security.

8.10 References

[1] WiMax, *IEEE 802.16a Standard and WiMAX Igniting Broadband Wireless Access*, White paper, www.wimaxforum.org/news/downloads/WiMAXWhitepaper.pdf.

[2] Z. Abichar, P. Yanlin, and J. M. Chang, "WiMax: the emergence of wireless broadband," *IT Professional*, vol. 8, 2006, pp. 44–48.

[3] J. P. Conti, "The long road to WiMAX [wireless MAN standard]," *IEE Review*, vol. 51, 2005, pp. 38–42.

[4] B. Rathgeb and C. Qiang, *Utilizing the IEEE 802.16 Standard for Homeland Security Applications, Proc. SPIE Defense and Security Symposiun: Technologies for Homeland Security and Law Enforcement*, Orlando, FL, USA, 2005.

[5] M. Donahoo and B. Steckler, *Emergency Mobile Wireless Networks, Proc. IEEE MILCOM*, Atlantic City, NJ, USA, 2005.

[6] *IEEE Standard for Local and Metropolitan Area Networks Part 16: Air Interface for Fixed Broadband Wireless Access Systems*, IEEE Std 802.16-2004 (Revision of IEEE Std 802.16-2001), 2004, pp. 0_1–857.

[7] C. Cicconetti, L. Lenzini, E. Mingozzi, and C. Eklund, "Quality of service support in IEEE 802.16 networks," *IEEE Network*, vol. 20, 2006, pp. 50–55.

[8] D. Johnston and J. Walker, "Overview of IEEE 802.16 security," *IEEE Security & Privacy*, vol. 2, 2004, pp. 40–48.

[9] *IEEE Standard for Local and Metropolitan Area Networks Part 16: Air Interface for Fixed and Mobile Broadband Wireless Access Systems Amendment 2: Physical and Medium Access Control Layers for Combined Fixed and Mobile Operation in Licensed Bands and Corrigendum 1*, IEEE Std 802.16e-2005 and IEEE Std 802.16-2004/Cor 1-2005 (Amendment and Corrigendum to IEEE Std 802.16-2004), 2006, pp. 0_1–822.

[10] L. Nuaymi, *WiMAX – Technology for Broadband Wireless Access*, John Wiley and Sons, New York, 2007.

[11] D. Pareek, *The Business of WiMAX*, John Wiley and Sons, New York, 2007.

Glossary

AODV Ad-hoc on-demand distance vector (routing) protocol

CA A certificate authority is an authority in a network that issues and manages security credentials and public keys for message encryption and decryption. As part of a public key infrastructure (PKI), a CA checks with a registration authority (RA) to verify information provided by the requestor of a digital certificate. If the RA verifies the requestor's information, the CA can then issue a certificate.

CRL A document maintained and published by a certification authority (CA) that lists certificates issued by the CA that are no longer valid.

DES A widely used method of data encryption using a private (secret) key that was judged so difficult to break by the US government that it was restricted for exportation to other countries. There are 72 000 000 000 000 000 (72 quadrillion) or more possible encryption keys that can be used. For each given message, the key is chosen at random from among this enormous number of keys. Like other private key cryptographic methods, both the sender and the receiver must know and use the same private key.

DH The Diffie–Hellman method for key agreement allows two hosts to create and share a secret key. Virtual private networks operating on the IPSec standard use the Diffie–Hellman method for key management. Key management in IPSec begins with the overall framework called the Internet security association and key management protocol (ISAKMP). That framework contains the Internet key exchange (IKE) protocol, which relies on yet another protocol known as OAKLEY and uses Diffie–Hellman.

DSDV Destination-sequenced distance-vector (routing)

DSR Dynamic source routing

GPS Global positioning system

ICMP The Internet control message protocol, an extension to the Internet protocol (IP). The ICMP supports packets containing error, control, and informational messages. The PING command, for example, uses ICMP to test an Internet connection.

IKE Internet key exchange is one implementation of ISAKMP. It is used to negotiate and exchange keying material between entities in the Internet. For example, IKE can be used to establish the IPsec security association. In IKE the Diffie–Hellman algorithm is used for the key exchange. Internet key exchange uses the same kind of two-phase SA as ISAKMP. In the first phase, IKE SA is created and in the second phase, keying information is changed and non-IKE SAs are established. The first phase may take place in one of the two modes. One of these protects the identity and the other doesn't.

IPSec A developing standard for security at the network or packet processing layer of network communication. The IPSec standard provides two choices of security service: authentication header (AH), which essentially allows authentication of the sender of data, and encapsulating security payload (ESP), which supports both authentication of the sender and encryption of data.

ISAKMP The internet security association and key management protocol is a key exchange independent framework for authentication, SA (security association) management, and establishment. It does not define the actual protocols to be used. The ISAKMP uses a two-phase approach in establishing SAs. In the first phase, the ISAKMP SA is established between the entities to protect the further negotiation traffic. In the second phase, the ISAKMP SA is used to establish other security protocol SAs, such as IPSec.

KeyNote Keynote is a simple and flexible trust management system that is designed for small and large internet based applications. It is fast enough to be used even in real time applications. It uses one easily human readable language to specify the policies and credentials.

MAC A bit string that is a function of both data (either plaintext or ciphertext) and a secret key, and that is attached to the data to allow data authentication. The function used to generate the message authentication code must be a one-way function.

MAC address A MAC address is used by the data link layer to deliver a frame to the destination node. Medium access control addresses are also called hardware addresses or NIC addresses, because this address is hard-coded into each NIC. Each type of network hardware has its own MAC addressing scheme. For example, the Ethernet uses 48 bit hardware addresses assigned by the vendor.

Mandatory access control (also MAC) Mandatory access control is an added security constraint enforced by a trusted operating system. It governs access to data, devices, or networks based on their sensitivity levels and, as the name implies, is mandated by the trusted operation system and cannot be changed or removed by users.

MD5 MD5 was developed by Professor Ronald L. Rivest. The MD5 algorithm takes as its input a message of arbitrary length and produces as its output a 128 bit "fingerprint" or "message digest" of the input. It is conjectured that it is computationally infeasible to produce two messages having the same message digest, or to produce any message having a given prespecified target message digest. The MD5 algorithm is intended for digital signature applications, where a large file must be "compressed" in a secure manner before being encrypted with a private (secret) key under a public-key cryptosystem, such as RSA. In essence, MD5 is a way to verify data integrity, and is much more reliable than checksum and many other commonly used methods.

MIB The set of variables or database that a gateway running network management protocols maintains. It defines variables needed by the SNMP protocol to monitor and control components in a network. Managers fetch or store into these variables.

PKI Public key infrastructure, a policy that establishes a secure method for exchanging information within an organization, industry, or country. It includes cryptography, the use of digital certificates and certificate authorities, and the system for managing the process. A PKI enables users of a basically unsecure public network, such as the Internet, to exchange data and money securely and privately through the use of a public and a private cryptographic key pair that is obtained and shared through a trusted authority.

PolicyMaker PolicyMaker is a trust management system. It is concerned with defining policies, credentials, and trust relationships. It uses a "safe" programming language to define the trust relationships, credentials, and policies. It is designed to be flexible enough to be used in large network applications and to integrate easily with the existing protocols.

Private key In cryptography, a private or secret key is an encryption and decryption key known only to the party or parties that exchange secret messages. In traditional secret key cryptography, a key would be shared by the communicators so that each could encrypt and decrypt messages. The risk in this system is that if either party loses the key or it is stolen, the system is broken. A more recent alternative is to use a combination of public and private keys. In this system, a public key is used together with a private key.

Public key A public key is a value provided by some designated authority as a key that, combined with a private key derived from the public key, can be used to encrypt and decrypt messages and digital signatures effectively. The use of combined public and private keys is known as asymmetric encryption. A system for using public keys is called a public key infrastructure (PKI).

QoS On the Internet and in other networks, QoS is the idea that transmission rates, error rates, and other characteristics can be measured, improved, and, to some extent, guaranteed in advance. Quality of Service is of particular concern for the continuous transmission of high-bandwidth video and multimedia information.

RSA One of the fundamental encryption algorithms or series of mathematical actions developed in 1977 by Ron Rivest, Adi Shamir, and Leonard Adleman. The RSA algorithm relies on the relative ease of finding large primes and the comparative difficulty of factoring integers for its security. The RSA algorithm is a public key cryptosystem for both encryption and authentication. It works as follows: take two large primes, p and q, and find their product $n = pq$; n is called the modulus. Choose a number, e, less than n and relatively prime to $(p-1)(q-1)$, which means that e and $(p-1)(q-1)$ have no common factors except 1. Find another number, d, such that $(ed-1)$ is divisible by $(p-1)(q-1)$. The values e and d are called the public and private exponents, respectively. The public key is the pair (n, e); the private key is (n, d). The factors p and q may be kept with the private key, or destroyed.

SA A security association (SA) is a relationship between two or more entities that describes how the entities will utilize security services to communicate securely. This relationship is represented by a set of information that can be considered a contract between the entities. The information must be agreed upon and shared between all the entities.

SASL The simple authentication and security layer (SASL) is a way to add authentication support to connection-based protocols. To use SASL, the protocol must include a command for identifying and authenticating the user to a server. The protocol may also include optional negotiation of a security layer for the subsequent protocol interactions. The command contains an argument that identifies the SASL mechanism to be used. If the server supports this mechanism, it initiates the authentication protocol exchange. This typically consists of changing challenge response pairs between the client and the server. These are specific to each protocol used. During the authentication protocol exchange the mechanism performs authentication, transmits an authentication identity and negotiates the use of a mechanism specific security layer. If a security layer is to be used, it is taken into use immediately and all the subsequent data exchanges are encrypted.

STS The station-to-station protocol (STS) is the three-pass variation of the basic Diffie–Hellman protocol. It allows the establishment of a shared secret key between two parties with mutual entitiy authentication and mutual explicit key authentication. The protocol also facilitates anonymity – the identities of A and B may be protected from eavesdroppers. The method employs digital signatures.

TLS The transport layer security (TLS) protocol. The TLS protocol provides communications privacy over the Internet. The protocol allows client–server applications to communicate in a way that is designed to prevent eavesdropping, tampering, or message forgery. The primary goal of the TLS protocol is to provide privacy and data integrity between two communicating applications. The protocol is composed of two layers: the TLS record protocol and the TLS handshake protocol.

WEP The 802.11 standard describes the communication that occurs in wireless local area networks (LANs). The wired equivalent privacy (WEP) algorithm is used to protect wireless communication from eavesdropping. A secondary function of WEP is to prevent unauthorized access to a wireless network; this function is not an explicit goal in the 802.11 standard, but it is frequently considered to be a feature of WEP. The WEP algorithm relies on a secret key that is shared between a mobile station (e.g., a laptop with a wireless Ethernet card) and an access point (i.e., a base station). The secret key is used to encrypt packets before they are transmitted, and an integrity check is used to ensure that packets are not modified in transit. The standard does not discuss how the shared key is established.

ZRP Zone routing protocol

Index

Printed in the United States
by Baker & Taylor Publisher Services